Technology Transfer,
Innovation, and
International Competitiveness

Technology Transfer, Innovation, and International Competitiveness

SHERMAN GEE

A Wiley-Interscience Publication

JOHN WILEY & SONS, New York • Chichester • Brisbane • Toronto

Library of Congress Cataloging in Publication Data:

Gee, Sherman.
    Technology transfer, innovation, and international
competitiveness.

    Bibliography: p.
    Includes index.
    1. Technology transfer. 2. Technological
innovations—United States. I. Title.
T174.3.G42   338.4'76   80-22786
ISBN 0-471-08468-9

Printed in the United States of America

10 9 8 7 6 5 4 3 2 1

To my wife Susanna
and daughters Darlene and Michelle

Preface

The old adage "the world is getting smaller" has never been more applicable. Modern technological advances such as supersonic air transports, satellite communications, or the ever-widening array of electronic products have caused distances to shrink since the time of our forefathers. Modern communications enable international trade to flourish and the sinews of a more interdependent world to develop. As a result, major events in one nation seldom occur in isolation from other nations. Political, economic, and military developments in one corner of the globe invariably exert an impact on other nations. This global interdependence is brought on by the progressive advance of modern technology, which has permitted the development of faster communications and closer international trade relationships as nations pursue increased industrial development and economic progress.

It is therefore not surprising that technology, economics, and national security have become increasingly intertwined in world affairs. Technology has emerged as the common denominator for nations seeking economic and military power, and the international transfer of technology has consequently assumed unprecedented importance in the aspirations of industrialized and developing countries alike. International technology flow since World War II has shifted the technological balance from one of unchallenged American leadership to one of virtual parity and high competitiveness with other Western industrial nations. The development of competitive technologies in other nations has reshaped America's foreign trade patterns and mutual security relationships with North Atlantic Treaty Organization (NATO) allies.

This book gives readers insight into the impact of technology on America's economic and security interests, and it underscores the importance of developing a balanced two-way flow of technology across our borders. It starts with a concise treatment of the relationship between technology and economics, followed by an examination of the importance of technology transfer in helping to expedite technological innovation. Methods of international technology transfer are outlined along with data on the extent of past

U.S. technology exports. Exports of U.S. technology have given rise to a stronger foreign technology base, increased trade competitiveness from foreign manufacturers, and changing U.S. relationships with NATO allies. The resulting economic pressures and higher NATO priorities dictate the need for policy and institutional reforms to improve cooperation with European allies and to revitalize American innovativeness and competitiveness. The book proceeds to describe how these ends can be achieved most practically by greater utilization of foreign technology and by working for greater benefits from foreign experiences in technological innovation and productivity enhancement.

Chapter 1 describes what is meant by technological change, innovation, and technology transfer, and establishes their interrelation with productivity and other factors of economic progress. Chapter 2 outlines the principal avenues of international technology transfer, traces the U.S. orientation towards technology exports, and describes the current movement towards a coherent U.S. technology export control policy. Chapter 3 focuses on the consequences of America's past emphasis on technology exports in terms of contributing to greater foreign technological competence, increased trade competitiveness, and changing U.S. military and trade relations with NATO countries. Chapter 4 articulates the need for the United States to utilize the technological resources of foreign countries more effectively in order to spur American innovativeness and trade competitiveness. Chapter 5 compares the institutional framework and the technological and economic policies in Japan and the Federal Republic of Germany for possible relevance and adaptation to the American scene. Chapter 6 assesses the current institutional make-up, public policies, and programs in the United States designed to foster increased American innovativeness and higher productivity growth. Chapter 7 suggests options for developing greater utilization of foreign technology and for improving conditions for technological innovation in the United States based on comparable foreign experiences.

This book will serve as a reference text for students of economics, business, engineering, public administration, and industrial management. It is also intended for industrial managers and government policymakers concerned with strengthening technological innovation in America. Researchers working in areas of research and development management, industrial innovation, and public policies will also find the book useful. It is my hope that the material will contribute to increased understanding of technology's pervasive impact on society and lead to better ways of harnessing and managing our nation's technological resources for socioeconomic progress.

SHERMAN GEE

Silver Spring, Maryland
January 1981

Contents

Technology Transfer,
Innovation, and
International Competitiveness

CHAPTER 1
Technology and the Economy

The conduct of world affairs today is increasingly swayed by the exercise of economic power rather than the traditional exploitation of military power. Historically, relations among nations have been dictated largely by their relative balance or imbalance in military strength. Today, however, we are seeing the emergence of economic power as an increasingly important factor in international relationships. Witness the pervasive consequences of the many-fold increase in oil prices initiated by the Organization of Petroleum Exporting Countries (OPEC). OPEC's ability to set the world price of oil imbues in the collective membership enormous power affecting the entire world economy that could never have been achieved by military means alone. Witness also the rapid, post World War II development of the Japanese economy into one that has surpassed the Soviet Union and rivals that of the United States. Japan's modern industrial machine supplies the world market with increasingly sophisticated manufactured goods and has extended the nation's sphere of influence world-wide, an achievement that proved elusive previously when military means were employed. One wonders also whether progress in negotiations between the United States and the Soviet Union on the Strategic Arms Limitation Treaty (SALT) and on issues relating to "detente" might not have been possible if it were not for the industrial strength and superior production technology and know-how of the United States. Normalization of relations between the United States and the People's Republic of China (PRC) might also not have occurred without the real stimulus of reciprocal trade benefits—modernization and industrialization of the PRC based on strong infusions of American capital and industrial technology.

Several factors account for the present-day emergence of economic relationships as increasingly important considerations in world affairs. First, the

1

contributions of the industrial sector of the U.S. economy to the Allied war effort during World War II is widely recognized. Without the extensive production base in industry already in place, the rapid change to an all-out war economy and the equipping of our armed forces with the necessary weapons, equipment, and supplies to face a well-armed and formidable foe would have been impossible. Our industrial strength coupled with a strong national resolve and commitment sustained the nation in overcoming early reverses and eventually led to the successful culmination of the war. Second, the world has witnessed the terribly destructive power of the atomic bomb, recognizing that what was seen represents only a small fraction of the destructive power of nuclear weapons currently deployed. The use of nuclear weapons in the next war will produce destruction of such an unimaginable and unprecedented scale on both sides of the conflict that there will be no real winners. The dire consequences of an all-out nuclear conflict will be so enormous that as a result war is looked upon by the world powers as the last-resort instrument of national policy. The world powers have replaced military combat with economic combat to achieve national objectives. Finally, economic well-being is recognized as the common thread underlying the hopes and aspirations of all nations irrespective of the political persuasions of government. Military strength, though necessary for national security, is but one potential instrument to achieve national objectives. More and more, military combat is being confined to settling local disputes among the smaller countries. It is therefore not surprising that economic considerations have begun to play an increasingly important role in international relationships.

A common denominator of both military and economic strength is technology, the technical means for doing things better, faster, more reliably, or cheaper. The contributions of technology in military conflicts are well known. For example, during World War II the use of radar in the European theater enabled both sides to extend their combat capabilities to nighttime and all-weather operations. Consider also the development of the atomic bomb, which resulted from a concerted effort by the United States to apply the scientific principles of nuclear fission to develop a practical weapon system. Technological developments such as these contributed immeasurably to the Allied war effort and helped bring a speedy conclusion to hostilities. Yet what is important is not technology per se, but the technological increment, change or improvement over existing alternatives that could translate into military (or economic) advantage. Although both sides during World War II used radar in military operations, only the British and the

Americans succeeded in developing radar systems that operate at microwave frequencies. Radar systems operating at these higher frequencies consequently were able to employ smaller antennas that were more suited for aircraft and ship operations. More importantly, microware frequencies allowed more accurate antiaircraft fire control, precision bombing, and more accurate ground control intercept for air defense. The development of the atomic bomb also reflects a singularly significant technological advance unequalled at the time, and its use is largely credited with the subsequent cessation of hostilities in the Pacific theater.

Technology, Productivity, and Economic Progress

The role of technology in the economic sphere is no less important, though perhaps less clear-cut, because of the many nontechnological determinants of economic progress. The widely used measure of economic progress is productivity, which is defined as output per unit or input. Productivity growth occurs when higher productive efficiencies are achieved. For the economy as a whole, higher productivity means better control of inflation, less unemployment, and a better quality of life in general. For the individual company, higher productivity means lower unit costs of products and increased market competitiveness. Similar to technological *change*, productivity *change* is the more meaningful descriptor of a nation's economic progress. The rate of productivity change is commonly used as an economic measure and is generally given in terms of annual percentage change.

National economic output is generally taken to be a nation's Gross National Product (GNP), which is the value of everything produced by all citizens including the goods and services produced outside the nation's boundaries. An alternative concept of economic output is the Gross Domestic Product (GDP), which is the value of all goods and services produced domestically by all workers including noncitizens.* National productivity then is given in terms of the GNP or GDP per employed person. At the industry level, however, output is measured by the number of production units manufactured or the value of services rendered. In this case productivity is typically given in terms of output units per labor hour.

*There are some who oppose the use of the GNP or the GDP as a measure of total economic output on the grounds that these measures fail to account for social costs—such as those arising from pollution control requirements, crime, drug abuse, and job health and safety—that have risen rapidly in recent years.

The input measure to productivity is more difficult to establish because of the diversity of the input factors that make up a modern, industrial economy. The labor investment in terms of worker-hours is most commonly employed as the economic input to productivity. However, where the use of labor productivity may have been adequate during the early stages of the Industrial Revolution when labor-intensive production prevailed in the main, it at best serves only as a rough measure of economic activity where production systems increasingly become more complex and capital intensive. Expanding the concept of economic input to include both labor and tangible capital, such as the value of available plant and equipment, would therefore make the productivity index more representative of actual production efficiency. Yet in the present age of increased automation and scarcer, more costly raw materials, contributions of technological progress and natural resources also constitute important elements of production. Additional elements affecting productivity are the education and training of the labor force, national economic policies, governmental regulations, management policies, working environment, and market conditions. Because the commonly used productivity index does not account for all these diverse factors, productivity should be looked on as only an approximate measure of economic activity.

It is clear then that technological change is but one of many factors affecting productivity. Nevertheless, technological change is widely viewed as the single most important contributor to productivity gains because it generally leads to more effective use of labor, capital, and natural resources—the other principal elements of production.*[1,2] Because the results of technological change are not easily measured, estimates of their contributions to productivity growth vary in the 40–70 percent range.[3] That is, roughly half of the productivity gains can be traced to technological advances that are reflected in improved manufacturing methods, materials, and machinery. By contrast about 15 percent of productivity gains result from capital formation, and approximately 12 percent comes from improved education and training of the labor force. Of the many input elements to productivity, then, technological advance appears to be the most effective avenue for fueling productivity growth. However, technological change is by no means synonymous with productivity improvement. To understand better the connection, let us look more closely at what is meant by technological change or progress.

*All source reference numbers in this text are sequential by chapter. (see Reference section)

Whereas science is the study of the universe to increase knowledge and understanding of it and how it functions, technology is the man-made creation that embodies the application of scientific knowledge and understanding. The output of science is usually publication of results. The output of technological activity is a product, process, technique, or material, the value of which is determined by the extent that it meets the user's need, objective, or requirement. Scientific and technological activity gives rise to the science-technology base that is the total accumulation of technical knowledge and resources developed.[4] The science-technology base is dynamic in that it is continually changing and growing as a result of scientific and technological advances, otherwise known as technological change.

Technological change is the incremental improvement or progress made in the science-technology base. It encompasses scientific advances, wider engineering know-how, availability of better materials, improved industrial processes including manufacturing methods, and more efficient design techniques. Technological change is the incremental upgrading of the existing pool of knowledge and accepted practices in the technical fields. It is the continuing process by which new discoveries and experiences are added to the science-technology base everyday. The change could be evolutionary in nature—highly specialized and limited in application—or it can be a truly revolutionary breakthrough (e.g., the invention of the transistor) leading the way to many additional incremental advances. More often than not, the technological change makes obsolete a portion of the existing science-technology base. Technological change then can, but does not have to, affect the bottom line of productivity improvement.

A popular notion is that in order to spur productivity growth by means of technological change, greater investments in research and development (R&D) is required. The rationale is that by increasing investment in R&D programs more R&D results will be generated that would be reflected in stimulating productivity. The record of R&D investment in the United States seemingly supports this notion as the onset of declining R&D budgets in the mid 1960s coincided with the start of a noticeable downward trend in national productivity growth. Furthermore, a look at the performance of major sectors of U.S. industry shows that the industries that devote a larger percent of sales to R&D are generally more profitable and competitive, whereas industries spending little on R&D are also the ones experiencing difficulties in meeting foreign competition. In the former category are the U.S. computer and electronics industries, which are the acknowledged leaders in the world market.[5] On the other side of the ledger are the steel and textile industries, where investments in R&D are low by comparison and where intense foreign competition has been encountered.

However, this coincidence of trends may be entirely illusory. Funding for R&D is a measure only of the effort to broaden the science-technology base. However, the relationship between the science-technology base and productivity improvement is not a close one. The path from R&D investment to productive output is typically a long and ardous one, lasting many years. Hence there is no assurance that the technical knowledge base resulting from R&D will be effectively applied towards greater economic output— new and better products and processes.

Evidence supporting the tenuous relationship between R&D and economic output can be found from several different sources. A recent investigation was made into the relationship between industrially funded R&D and productivity growth in the manufacturing sectors—chemicals, metals, electrical and nonelectrical machinery, and instruments—in the 12 industrial nations of the Organization for Economic Cooperation and Development. (OECD) for the period 1963-1973.[3] The analysis was performed for each of the four industries as well as for the total manufacturing sector across all 12 countries. No clear correlation between industrial R&D intensity and productivity was found. If we also examine a measure of the basic research output of a country, namely, the number of Nobel prizes awarded per unit population, in relation to productivity improvement, a strong inverse correlation in fact appears. That is, countries with the most Nobel prize laureates over the 1961-1976 period are, in descending order, the United Kingdom, United States, West Germany, and Japan. However, the ranking of nations exhibiting the highest productivity growth during the same period is almost exactly the opposite: Japan, West Germany, United Kingdom, and the United States. The evidence thus strongly suggests that R&D and productive output are not in fact closely coupled.

Lending further credence to this view is evidence that corporate profitability does not correlate well with the introduction of significant new products despite available data showing that corporate profitability correlates well with R&D investment as a percent of sales.[6] The implication, therefore, is that R&D investment is not closely correlated with the introduction of new products. If R&D investment indeed is linked closely with productivity growth, one would also expect good correlation between R&D investment and the introduction of new products into the marketplace inasmuch as economic growth is generally dependent on the ability to generate a continuing stream of new or improved products and services into commerce. That this is apparently not the case is further indication that the (innovation) process that translates R&D results to productive output is dependent on many complex variables, only one of which is R&D investment. Certainly there is no prima facie reason that profitability should correlate well with numbers of new products because it is entirely possible that no market acceptance is generated by any of the products introduced. Also, high profitability can result from only a few established products that enjoy long production runs and lasting market demand. Looking at it from another standpoint, profitability is generally associated with established products already well down their respective life cycles, whereas new products have hardly begun to build their own profitability track record and consequently have not impacted on the company's profitability.

Consider also that R&D is a contributor to technological change but is not necessarily its principal determinant. Technological change can occur without R&D for example, from serendipitous discoveries and personal experiences. Furthermore, technological change does not automatically translate into productivity growth but is only one contributor to a complex process by which commercial innovations are introduced into the economy. In fact, as will be evident later, R&D itself is found not to be a very significant factor in this process. Therefore, it is not surprising that no clear correlative relationship can be found between R&D investment and productivity growth.

Nevertheless, technological change will continue to be an important factor in fueling productivity improvement. Consider, for example, the present trend of higher energy prices. There is little doubt that higher energy prices, triggered by OPEC's dramatic raising of the world price for oil, will remain with us into the twenty-first century until alternative sources such as nuclear and solar energy can be fully brought on stream. What effect will higher energy prices have on productivity? Investigations into this question have centered on the energy relationships with the productive inputs—labor and capital. The general conclusions drawn are that energy and labor are substitutable inputs to productivity, while energy and capital are complementary.[7] That is, higher energy prices will increase the demand for labor as a substitute, while more expensive energy will dampen the demand for both energy and capital. However, these relationships do not necessarily prevail in all situations. If labor wage rates rise faster than energy prices, then energy and capital will continue to be substituted for labor. On the other hand, higher energy prices could force more capital investment in energy conservation measures. Nevertheless, the net economy-wide effect of higher energy prices is to increase demand for energy-saving labor and to reduce the demand for investment capital. The resultant effect would be to slow down productivity growth because the substitution of capital for labor in the production system, which historically has been one of the more important stimulants for productivity growth, will be dampened. As a result, contributions of technological change for productivity improvement becomes even more important because of the promise of developing labor-saving innovations and more energy efficient plants and equipment that helps to temper high labor demand and to encourage capital investment.

Looking at other unmistakable socioeconomic trends apparent today and their effect on the other input factors to productivity further supports the notion that technological change provides the most promising avenue for higher productivity gains. The continuing inflationary spiral coupled with the increased scarcity of nonrenewable resources will cause the price of raw materials to continue its upward trend at a fast pace, thereby limiting the potential for raw materials to contribute significantly to a higher rate of productivity increase. The present trend towards more government regulations for protecting the environment and the safety and health of the worker on the job will add to the social costs associated with the free-market economy and consequently will tend to lower future productivity gains. Current United States efforts to control more effectively the export of critical technologies and technological products in effect narrows the potential market for many technology-intensive products. As a result, the potential for greater economies of scale and the attendant increases in productive efficiencies possible from serving a world market are visibly constrained. All these present trends consequently act to inhibit greater productivity gains. At the same time they serve to underscore the importance of technological change for spurring productivity improvement.

Though technological progress may become even more important for achieving higher productivity in the future, it must still be recognized that it does not automatically result in improved productivity. Only to the extent that technological advance is coupled to the process by which new and improved goods and services are introduced into the commercial market does technology contribute to productivity. The process by which technological advance is translated into competitive goods and services—technological innovation—is therefore a prime determinant of the effectiveness by which technological progress is translated into higher productive output in the national economy.

Technological Innovation

Innovation is the process of taking an idea, invention, or recognition of a market need and developing a useful product, technique, or service to the point where it gains initial commercial acceptance. Innovation also encompasses the refinement and improvement of an existing product, technique, or service towards commercial acceptance. It is characterized by the creative assimilation of seemingly disconnected and diverse elements into the development and user acceptance of a new product, process, or technique. The innovation process is generally long and tortuous, spanning many years from initial conception to the point of first realization or user acceptance. During the innovation period, external conditions—such as the changing market structure, competition, government regulations, management support, and the availability of capital—that are normally beyond the control of the innovator act to change the risk-reward outlook to the extent that more often than not the innovation must be prematurely terminated. That is, serendipidity and luck play a significant role in the success or failure of an innovation. As a result, the innovation process is evolutionary in nature and is not subject to strict management and control.

Innovation does not necessarily depend on technology. It can occur entirely in the arts, in education, or in a social context. For example, a teacher in a school classroom may innovate by simply changing the instructional plan to include novel visual displays as a teaching aid to increase student learning and retention. Or, as a second example, innovation in the arts may involve the pioneering of new artforms, techniques, and interpretations that are expressive of the full range of human emotions and the grandeur of nature. The scope of innovations may be quite limited, requiring only a few days to be adopted, or they may be highly complex with time scales extending over many years.

Technological innovation refers to innovation that draws on the available science-technology base. As such it acts to link existing technological resources to economic output. Although R&D is often an important element in the process, it is insufficient by itself because the successful realization of an innovation depends on a host of other factors, such as the market environment and management disposition towards continuing to pursue the innovation. Application of technology and R&D can occur at any point during the series of activities that make up the innovation process. These activities may be considered in three distinct stages.[8]

The first stage is initiated at the first conception of the idea, invention, or recognition of a market need. It involves the evaluation of available knowledge and alternative concepts that seem promising in providing a technically and economically feasible means to implement the idea, make the invention, or fill the market need. If available knowledge is insufficient, research and development may be conducted to add to the technology in the technology base, to test certain hypotheses, or to verify feasibility. Of course, early in the first phase management must be convinced and must authorize the initiation and pursuit of the innovation before further progress can be made. If no serious impediments are encountered at this stage, then the second stage may proceed during which more detailed technical, cost, and market analyses are conducted, further technical development performed, problems solved, and prototypes developed and tested. At this stage if technical feasibility and economic potential continue to look promising, then the third stage is initiated. This final stage involves final product engineering design, test, manufacture, marketing, and introduction into the market, at which point first realization occurs and the innovation process is substantially complete. Wider market penetration, further technological improvements and adaptation of the innovation to new market needs typically take place after first realization of the innovation.

The latter stage of the technological innovation process is characterized by entrepreneurial forms of activity spanning product engineering, manufacturing, and marketing. Because this stage of the process includes possibly large capital investments in plant, equipment, materials, and human resources as well as significant outlays in market development, setting up distribution networks, advertising and promotion, test marketing of the product, and the like, the decision to proceed is made only after thorough and detailed deliberation by management. Typically, costs associated with the third phase of the innovation process are at least 10 times greater than costs expended in the first phase of the process. Still, about 30 percent of innovations enter this most expensive third stage, which is certainly too high considering that only about 2 percent eventually emerge from the innovation pipeline as successful realizations.[9]

The entire innovation process then is built up from many distinct links forming a long chain of events. Failure of any of the links in the chain will cause the innovation to fail. Crucial to the process, of course, are the efforts of dedicated and imaginative people with entrepreneurial qualities to see the innovation through to commercial realization. Without these product champions, an innovation will have little chance to succeed.

R&D in Innovation

R&D is often required to respond to an identified need or problem arising from a specific innovation. R&D, in this sense, helps to establish technical feasibility or to test cardinal hypotheses and assumptions related to the innovation. It serves to expedite an existing innovation rather than to initiate new ones. The supposition is often made that R&D is a prominent source of new technological innovations. However, evidence to date in fact suggests that R&D plays only a minor role in helping to stimulate new innovations.

Numerous studies have been performed in order to increase our understanding of the innovation process and, in particular, to identify those factors most commonly associated with successful innovations. The picture that emerges is one that points to the clear importance of being attuned to the market environment. A major conclusion drawn from the studies is that roughly three out of every four successful innovations are stimulated from need recognition, while the remainder are initiated from the availability of technical opportunitites.[10] That is, demand-pull rather than technology-push is the more important stimulus in most cases of successful technological innovations. This is not to imply that idea conception, invention, or recognition of technical opportunity are unimportant in innovation.[8] On the contrary, they could very well play crucial roles throughout the different stages of the innovation process to the extent that the innovation could otherwise fail without them. The point is that need recognition is the most important factor in *starting* an innovation, while the other factors could be important at any stage of the innovation process itself.

The telephone is an example of an important innovation that was stimulated primarily by recognition of the market need. The instrument was invented and introduced to the world as a result of experiments originally conducted by Alexander Graham Bell in pursuit of a "harmonic telegraph."[11] He recognized the wide commercial potential for a telegraph system that was capable of transmitting musical tones, and thus of transmitting several messages simultaneously on a single wire instead of only patterns of dots and dashes. That Bell eventually invented the telephone instead, a much more important discovery than originally intended, is illustrative of the evolutionary nature of the creative process. It is important to note, however, that he originally had no knowledge of electricity, but rather was trained in the physiology of human speech. Bell had to acquire the basics of electricity in the course of pursuing his experiments, and he relied on an able young assistant for competence in mechanical matters. Hence the stimulus for Bell came not so much from the presence of a technical opportunity, but from a clear recognition of the telephone's potentially vast commercial market.

The difficulties associated with realizing innovations stimulated by technical opportunities are illustrated by the experience of the laser industry.[12] Development of the laser in the early 1960s was generally expected to be the precursor of a $1 billion laser industry by 1970. However, despite considerable attention devoted to developing commercial laser applications, civilian sales of lasers, components, and accessories by 1970 totaled less than $45 million. This failure to live up to early commercial expectations is attributed primarily to inadequate attention by the technical entrepreneurs to market considerations.

These examples point to the increased difficulties encountered when innovations are developed from predominantly technological considerations without adequate knowledge of the potential market demand. Innovations created in response to recognized needs are necessarily market oriented instead of technology oriented. Consequently, they tend to gain more rapid acceptance in the user community. These innovations are normally pursued with an eye toward market introduction at the earliest possible date because of the transitory nature of market demand, the rapidity with which technological obsolescence develops, and the opportunity to scoop the competition. As a result, much of the early effort during the innovation process concerns the search for relevant information and the application of existing technology rather than the conduct of R&D. Indeed, the necessity for performing R&D may be viewed by the innovator as being more of a hindrance because of the inherent uncertainties in scheduling, costs, and results associated with any R&D effort.

It is not surprising then that R&D generally is not the overriding factor in technological innovation. Technological opportunities account for only a small proportion of successful innovations. Studies of successful innovations in both the military and civilian sectors reveal that research results initiated innovations in only about 5 percent of the cases studied.[13]

R&D activity for technical problem-solving during the course of an innovation also appears to play only a minor role. A Defense Department study on the role of research in weapon systems development during the 1945–1962 period found that only 9 percent of over 700 significant scientific and technological events associated with the development of 20 major weapon systems could be identified as resulting from basic or applied research.[14] It has also been reported that nearly 80 percent of the information used in solving technical problems during innovation were widely available and did not require R&D.[15] The major sources of technical information were printed materials and personal contacts. Nevertheless, R&D does contribute to innovation in a critical but indirect way by means of education and general information for the innovator.

The typical problem-solver generally relies on his own knowledge and experience as much as possible before turning to alternative information sources. His breadth of knowledge and experience are important factors affecting the manner and dispatch by which technical problems are resolved. In this way the eventual outcome of an innovation is strongly determined by the innovator's background and educational level. His educational level also appears to affect the patterns of information flow employed. External sources of information are relied on more heavily by individuals with a university education as opposed to those with less formal education, who depend more on personal information.[15]

The low significance of R&D in innovation is also reflected in the differences in characteristics and motivations of the innovator and the researcher. The innovator typically exhibits entrepreneurial qualities requiring a breadth of knowledge in different fields, whereas the researcher possesses in depth knowledge in a specialized field.[16] The innovator uses available knowledge as much as possible, while the researcher pursues the generation of new knowledge. The innovator generally exhibits a high degree of creative ability. In contrast, creativity appears to have no direct relationship with manifest research competence.[17] The innovator is strongly motivated with the promise of commercial success. On the other hand, the ability to pursue one's research interests coupled with the peer recognition associated with significant research achievements are the prime motivational forces for the researcher. These distinctions, however, become increasingly less clear for those participating more in the applied research and exploratory development facets of the R&D process. Nevertheless, they serve to illustrate the differences in the basic behavioral characteristics and motivations of innovators and researchers that contribute to the low impact of R&D on the innovation process.

Stimulating Innovation

Many studies have been conducted looking into the important factors that seem to characterize successful innovations. One such study looked into the differences between success and failure in innovation by investigating cardinal events surrounding some 70 innovations equally divided between successes and failures.[16] The 35 success/failure pairs were selected from two industries, chemicals and scientific instruments, to minimize influences arising from differences between industries. Innovations that realized commercial success were found to be superior in the following areas: (1) innovation-inclined management, (2) understanding user needs, (3) marketing and sales performance, (4) efficiency of development, and (5) effectiveness of communications. The background, status, and personal attributes of the manager having direct responsibility for an innovation were found to be crucial to whether the innovation ultimately succeeds or fails. Sensitivity to the external environment is important in terms of understanding and being responsive to user needs and in promoting the innovation product with strong marketing and sales efforts. Efficiency of development accounts for the necessity for maintaining high standards of product quality for achieving commercial success. The maintenance of effective communications both internal and external to the organization is necessary to help assure the timeliness of information flow. With the major informational task being that of searching for and applying existing knowledge, the existence of effective communication channels often prove to be the difference between success or failure in innovation. Additional factors found to be important in other studies include after-sales service and user education, the presence of strong individuals committed to expediting the innovation process, and a corporate-wide identity and support for the particular innovation.[18]

The extent to which these factors appear in any organization are, of course, subject to the prevailing management policies, organizational goals, individual attitudes, and the general economic environment. If there is one area that approaches universality in importance, it is the need to establish effective communications. The importance of effective communications is magnified by the realization that it constitutes the one factor with the most direct potential for improving the utilization of R&D results in innovation. With easier access to R&D results and a stronger inclination by technical innovators to seek the information, freer flow of R&D results could substantially improve the level of R&D contributions to the innovation process.

The time period for innovation typically spans 3–30 years or more. The time span for technological innovation is important because it directly affects the rate at which innovations are produced. Developments tending to shorten the innovation process will help generate a higher rate of technological innovation. This is desired for increased competitiveness and timely response to market opportunitites. However, the entire process is subject to many unpredictable, complex external forces, which are generally beyond the control of the innovator and often determine whether the innovation succeeds or fails. The situation is aptly described by the following passages from President Nixon's Science and Technology Message to Congress delivered on March 16, 1972:

> In the first place, we must always be aware that the mere act of scientific discovery alone is not enough. Even the most important breakthrough will have little impact on our lives unless it is put to use—and putting an idea to use is a far more complex process than has often been appreciated. To accomplish this transformation, we must combine the genius of invention with the skills of entrepreneurship, management, marketing, and finance.

> Secondly, we must see that the environment for technological innovation is a favorable one. In some cases, excessive regulation, inadequate incentives, and other barriers to innovation have worked to discourage and even to impede the entrepreuneurial spirit. We need to do a better job of determining the extent to which such conditions exist, their underlying causes, and the best ways of dealing with them.

That is, the existence of a large science-technology base is useless unless it is employed to strengthen our economy and improve the quality of life. To do so requires creating an environment conducive to technological innovation and overcoming or minimizing the barriers that act to impede the process.

The primary barrier to technological innovation relates to marketing.[9,19] More specifically, uncertainty of market information, market fragmentation limiting market size, and difficulties in identifying and developing new market areas are considered the main impediments. Poor management or over-management is the next most common reason for failures in innovation.

Included also is the tendency on the part of too many companies to emphasize short-term, low risk projects rather than the more venturesome innovations having the promise of higher commercial payoff. Marketing and management problems account for roughly half of the innovations which fail. Other reasons for failures in innovation arise from lack of capital, particularly new venture capital for high-technology businesses, and governmental policies and practices relating to patent, antitrust, and other regulatory matters. Interestingly, technological factors such as the availability and dependability of technical information are not considered important impediments. The picture that emerges is one in which the developing science-technology base is adequate to fuel technological innovation. The major difficulties are nontechnological and are related to marketing, management, capital, and government regulations. It would appear that these obstacles are the same as those found in the normal business practices of industrial firms and therefore can be expected to be important also in innovations that do not involve the science-technology base.

The growing influence of government is especially felt in the regulatory area. With the rising social consciousness of recent years, industrial firms must devote more and more attention to meeting government regulatory requirements for environmental protection, occupational safety and health, and for a host of other national concerns. There are over 85 federal agencies regulating business and industry at a cost of more than $100 billion, of which approximately $60 billion is estimated to be spent on unnecessary regulations. While the objectives of government regulatory activities are beyond reproach, such as cleaner air and water and removing health hazards from the workplace, the costs of regulation compete for funds needed for capital improvement and plant expansion. This is a major reason for the slower rate of capital investment in the United States in recent years.

The concern also is that overregulation will exact too high a price in terms of unnecessarily high costs being passed on to the consumer.[1,20] Unnecessarily stringent standards for protecting the environment will require the installation of overly expensive pollution control equipment, resulting in excessive energy consumption, unnecessarily high production unit costs, reduced productivity, and higher inflation rate. Chase Econometrics estimated that environmental regulations added a half percentage point to the high inflation rate in 1978. Furthermore, the costs required to meet government regulatory requirements are not small. Pollution control expenditures alone are expected to total nearly $650 billion in the 1980s. Taking into account expenditures required to meet government standards in the other regulated areas, it is clear why government regulations are viewed by business and industry primarily as a major disincentive for industrial innovation.

This view does not appear to be shared by the general public, however. At the time that regulatory rules were formulated to assure that adequate product testing was performed to assure environmental and consumer safety, the public consensus was in favor of conservatism over cost consciousness.[21] That this sentiment has remained unchanged is borne out by two separate public polls conducted in 1978.[22] In response to whether environmental standards should be relaxed to conserve energy or to aid the economy, public opinion was found to be strongly opposed to loosening environmental standards by a three to one margin. In another survey conducted in 1977, 68 percent of those polled favored higher prices and taxes to protect the environment, and 55 percent felt the environment should be protected *at all costs*. Clearly, the public assigns less importance to the considerable costs associated with environmental cleanup and protection than do business and industry.

It should be recognized too that whereas loosening stringent standards will help lessen inflation, it will not necessarily quicken the pace of innovation materially. Regulatory roadblocks to innovations may be moderated, however, with governmental assistance in interpreting the regulatory requirements to determine applicability to specific industry situations and also in providing advice on how the requirements can best be met.[23] Additionally, the government can try to minimize uncertainties and inconsistencies in existing regulatory requirements. Business views a stable economic and regulatory climate where unknowns are minimized as the key stimulant for industrial innovation. Uncertainty, on the other hand, may deter a company from investing in new capital equipment, starting a new business venture, forecasting future markets, or complying with a regulation because of the tendency for requirements and standards to change. A stable regulatory climate then will allow normal business planning activities to proceed; without these activities, the pace of innovation will surely suffer.

Recognizing the many factors which act to hamper innovation should therefore help to identify measures to strengthen the process. Much attention has been focused on the need for federal intervention to provide incentives for encouraging innovation, such as increased federal support for commercially oriented R&D, loan guarantees for capital investment, and tax incentives to encourage industries to invest more in R&D and capital equipment. These incentives are based on the presumption that normal free-market forces are inadequate and that greater federal assistance is therefore required. However, no consensus has been reached on the desired form of government incentives. There is inadequate assurance that governmental measures, such as more federal funds for commercially relevant R&D and formulation of tax and capital incentive programs, will indeed address the root causes of the slower growth in the U.S. economy. R&D investment is, after all, only an input measure that is at best only indirectly related to productivity growth. The lack of correlation in patterns of R&D funding with productivity growth in the manufacturing sectors of the OECD countries bears this out. Furthermore, the federal incentives considered are macroscopic measures that are generally remote from significant individual impact. That is, monetary incentives to the company will not necessarily translate down to more creativity, inventiveness, dedication, and entrepreneurship in the individual, and these traits are the essence of the innovation process. Compounding the situation is the perceived inability to exploit fully the scientific knowledge and technical know-how already available.

The situation is not entirely bleak. The essential issue is to forge a stronger linkage between the growing science-technology base and economic progress by strengthening the process for technological innovation. Better access to technical know-how and more effective utilization of the science-technology base for spurring innovation and productivity growth must be developed. Conscious efforts are required to put technology to work in new applications and circumstances originally unforeseen during its first development. A system for the transfer of technology is needed to reap the full potential of the available science-technology base.

Importance of Technology Transfer

Simply put, technology transfer is the application of technology to a new use or user.[8] It is the process by which technology developed for one purpose is employed either in a different application or by a new user. The activity involves principally the increased utilization of the existing science-technology base in new areas of application as opposed to its expansion by means of further R&D. It provides the means by which elements of the existing science-technology base can be more closely coupled to the innovation process in order to spur productivity growth. The time span for the transfer can be quite short—a matter of days in cases where the technology transferred may be directly applied in its existing form to the new environment—or the process can extend to a considerable number of years in cases where extensive modification of the technology, redesign, or adaptive engineering is required to make the technology fit its new role.

A variety of means is available to transfer technology depending on the nature of the technology and the specific circumstances prevailing in each case. Methods range from licensing agreements, joint ventures, and turnkey factories to technical consulting, product sales, and trade exhibits. No single method is appropriate for all situations. The effectiveness of the different approaches differs in terms of the ability of the technology recipient to learn and to acquire increased technological know-how. Generally, those methods that involve considerable person-to-person contact and some measure of education and training are considered the more effective approaches. Successful technology transfer therefore hinges, to a large extent, on whether effective communications is developed between the principal parties. Effective communications, however, is a necessary but not sufficient condition to assure success in technology transfer. Many other technical, economic, and social factors also have a bearing on the eventual outcome.

Technology transfer is not new but has been in existence for a long time. In the thirteenth century Marco Polo helped introduce to the Western world Chinese inventions such as the compass, papermaking, printing, and the use of coal for fuel. It has been only in recent years, however, that a concerted and systemized effort has been made to transfer technology as a means to stretch the R&D investment dollar and to develop greater utilization of the existing science-technology base in order to generate greater economic impetus. Technology transfer offers the opportunity to obtain a greater return from past investments in R&D, but is not an end in itself. Its importance lies in its ability to stimulate and strengthen the innovation process. As such it provides the means by which the rate of technological innovations might be increased. To understand the importance of technology transfer in the innovation process, it is necessary to look more closely at how technology transfer interrelates with innovation.

Let us depict the innovation process as a simple time line beginning at the point of first conception and ending when the innovation is first realized and introduced into the market.[8] The time span represented by the time line is the innovation period encompassing the myriad of activities that make up the innovation process. The use of the time line, of course, does not mean that all innovations proceed to successful realization. It does, however, provide a convenient means for characterizing the successful innovations. Innovations that draw on the science-technology base, that is, technological innovations, may be both recipients of technology and sources of technology to expedite other innovations.

The situation may be illustrated as shown in Figure 1-1. The time lines A and B represent two different innovations where C and R are respectively, the points of conception and realization. Technology arising from innovation A is transferred to innovation B as depicted by the dashed transfer lines 1 and 2. In the case of transfer 1, the product of innovation A may be incorporated in innovation B as a process innovation or as an aid in demonstrating feasibility. Transfer 2 shows that technological by-products of innovation A may find application in innovation B, for example, in providing for improved prototype testing or more efficient manufacturing. The technology transferred from innovation A lowers the technical risks associated with innovation B, avoids the need to "reinvent the wheel," and expedites innovation B to successful completion.

The net effect on innovation B is likely to be a distinct shortening of the innovation time period. Shortening the innovation time period means increasing the speed and efficiency with which technological innovations are introduced into the economy. This is a valuable competitive advantage because it could lead to capturing a larger share of the market or to profiting from a temporary market demand. The ability to innovate rapidly, of course, makes possible a higher rate of technological innovation, which is needed to fuel productivity growth.

Innovation A also stands to benefit from the technology transfer, as represented by path 1 of Figure 1-1. Introducing the product of innovation A into innovation B effectively promotes the diffusion of the product into new

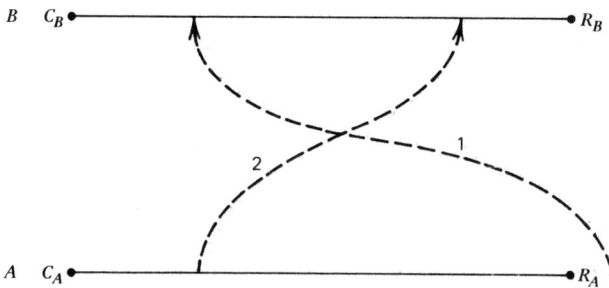

Figure 1-1   Technology transfer interaction with innovation. (Source: Reference 1-8) (Reference 1-8 is Reference 8 in Chapter 1.)

market areas. The resulting increased demand for the product generates greater economies of scale through higher volume production. Unit costs are therefore lowered, enhancing the product's potential for penetrating further into new market areas previously inaccessible because of the formerly higher product price. The new market areas generate increased product demand, making possible further economies of scale, lower unit costs, and so forth. Another obvious advantage arising from successful penetration into new market areas is that experience gained should produce greater knowledge of market characteristics and trends; also it should increase the likelihood that initial market penetration can be further expanded through the introduction of additional new products. Therefore, the introduction of product *A* into the new market area represented by *B* could well be the bridgehead needed for appreciable follow-up business expansion.

We have seen how both innovations *A* and *B* could benefit from the technologies transferred as shown in Figure 1-1. It also should be clear that *A* can benefit in the same fashion as *B*, and vice versa, since the technology flow paths 1 and 2 as shown are completely general in that they may also flow from innovation *B* to *A*.

An innovation of course can draw from more than one source of technology. This situation is illustrated in Figure 1-2, where innovation *D* is benefiting from technologies obtained from innovations *A*, *B*, and *C*. The invention of the turbojet engine was itself derived from two technologies—the gas

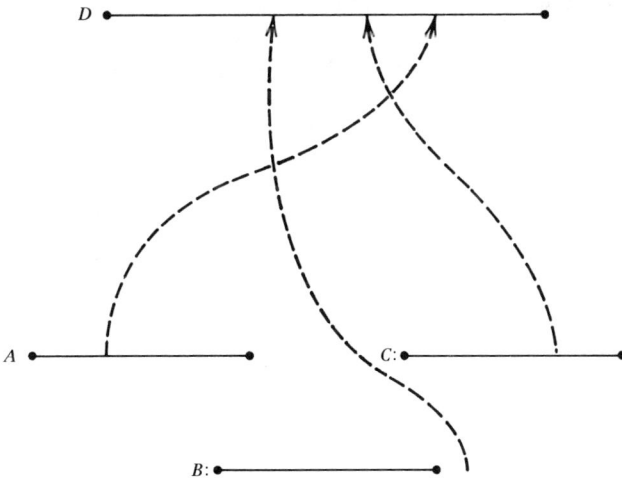

Figure 1-2   Technology transfer from multiple sources. (Source: Reference 1-8)

turbine and the jet propulsion principle. A more recent example of an innovation benefiting from multiple sources of technology is the train antiderailment sensor developed by the Naval Surface Weapons Center for the Federal Railroad Administration. The device, or hot-box detector, senses the temperature of the train wheelbearings and is designed to trigger the braking system to prevent incipient derailment when the wheelbearing temperature rises above a preset threshold. The design of the hot-box detector benefited from two previous Navy innovations, the fuzing technology previously developed for a widely used ordnance device and the nickel-titanium intermetallic compound, Nitinol, discovered as part of a separate effort to find new materials for reentry vehicle applications. The combination of these two technologies made possible the design, fabrication, and feasibility testing of the hot-box detector in little more than a year. Although the hot-box detector has not yet been commercially realized, the availability of the fuzing and Nitinol technologies undoubtedly spurred the substantial progress made during the first year, suggesting that multiple sources of technologies indeed produce a measurable shortening of the innovation time period.

Further evidence of the positive influence that accessing multiple sources of technology might have on the innovation process is found from data on 500 significant innovations introduced into the marketplace during the 1953–1973 period.[24,25] The 500 innovations were selected from an original field of 1300 innovations by an international panel of experts on the basis of their technological, socioeconomic, and political significance. The majority of the data dealt with American innovations. We are specifically interested in the innovation time span and the numbers of technology sources used in the 258 American innovations. Table 1-1 summarizes this data.

The table shows that innovations using from one to three technology sources are about equally likely to be found. Innovations depending on four

Table 1-1   Average Innovation Period for U.S. Innovations According to the Number of Technology Sources

	Number of Technology Sources				
	1	2	3	4	5
Number of innovations	65	67	68	52	6
Average innovation period (years)	7.4	8.0	7.8	4.9	5

Source: Reference 1-25.

or more technology sources become distinctly less likely. Furthermore, the average innovation period appears to be the longest for innovations involving two and three sources, and shortest for those cases involving four sources. However, the two-source data includes an innovation that required an unusually long 82 years to complete, while the three source data includes two innovations each of which took 63 years to complete. These time spans are much longer than the average for their respective groups, and they represent dispersions about the mean much greater than for the rest of the data. After removing these nonrepresentative cases from consideration, the average innovation time periods for the two-and three-source data become 6.8 and 6.1 years, respectively.

These adjusted data are used in Figure 1-3, which depicts the dependence of average innovation time on numbers of technology sources. The figure indicates a general inverse relationship. However, the paucity of five-source data necessitates regarding the corresponding data point as highly tentative, as represented by the dashed lines in the figure. Nevertheless, the inverse relationship indicates that using more sources of technology on balance will tend to shorten the innovation time period.

For this reason the innovator should be sensitive to external technological developments. Of course the ability to use outside sources of technology depends on effective and timely communications. Technology transfer nec-

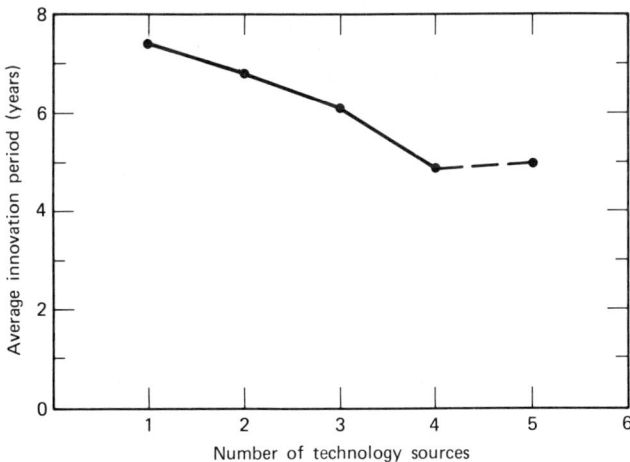

Figure 1-3   Average innovation period in the United States as a function of the number of technology sources. (Source: Reference 1-25)

essarily must be preceded by the flow of appropriate information. As mentioned earlier, the communication mode considered most effective for fostering technology transfer is person-to-person contact. Direct interpersonal communication enhances understanding and commitment to action, whereas reading printed matter generally creates awareness at best. Direct personal interaction also greatly facilitates the generation of new ideas and the pursuit of unforeseen innovations. Thus the process of transferring technology not only expedites an innovation but could also open up additional avenues of investigation leading to new, unplanned innovations.

This situation is illustrated in Figure 1-4, where technology transfer path 1 stimulates the conception C_{B2} of a new innovation and transfer path 2 could similarly lead to the subsequent conception C_{B3} of a second unplanned innovation.[8] The introduction of the turbojet engine in commercial aircraft in the United States evolved from just such a situation. The turbojet engine, first developed and tested in England in 1937, was transferred to the United States principally for military development just after the beginning of World War II. It found subsequent application after the war in the civil air transport industry where it was introduced into com-

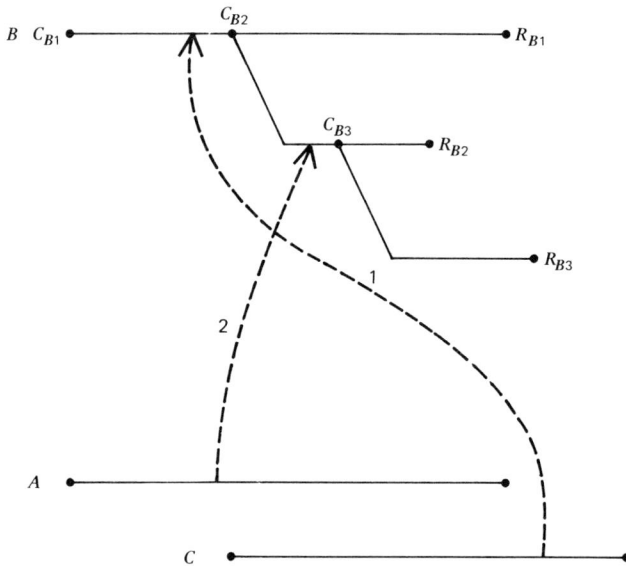

Figure 1-4   Multiple innovations stimulated by technology transfer. (Source: Reference 1-8)

merce with the first flights of the highly successful Boeing 707 aircraft. The turboject engine has been further tested in yet another potential application—fog dispersal at airports. Thus the transfer of turbojet engine technology from England originally planned for U.S. military aircraft development helped spawn two originally unplanned innovations—the Boeing 707 aircraft and a potential fog dispersal mechanism.

Another innovation that helped spawn an entirely new technology area, and subsequently a new industry, is the klystron tube.[26] The klystron is a source of microwave energy invented in the United States in 1937 just prior to World War II in direct response to the need to detect aircraft attacking cities in bad weather or in darkness. The short wavelengths of the microwave energy permitted accurate radar tracking of airborne targets unobservable to the naked eye. The technology was subsequently acquired and adapted by the British for airborne radar applications. Lightweight klystron radar receivers on Royal Air Force aircraft increased the effectiveness of their nighttime operations and is credited with helping win the Battle of Britain. The klystron was an important factor in developing many other nonmilitary applications—satellite communications, commercial air navigation, and medical and nuclear instrumentation. It also helped stimulate considerable research and development in microwave devices, instruments, and techniques during and after the war years.

Technology transfer thus plays an important role in fostering technological innovation. It provides opportunities for employing proven technologies rather than relying solely on internal R&D to satisfy technological needs. The use of external sources of technology tends to shorten the innovation time and could also mean the difference between success or failure of the innovation. Technology transfer also holds the promise of stimulating new ideas and new innovations. The cumulative effect is to make more effective use of the science-technology base and to produce a higher rate of technological innovation.

The Technoeconomic Interface

As we have seen, the interface between the science-technology base and economic progress is a complex one buffeted by strong social, economic, and political currents. Not everyone is able to work effectively within the technoeconomic interface. Thomas A. Edison, famous for his inventive ability, for example, in pioneering the electric lighting system, is often considered the

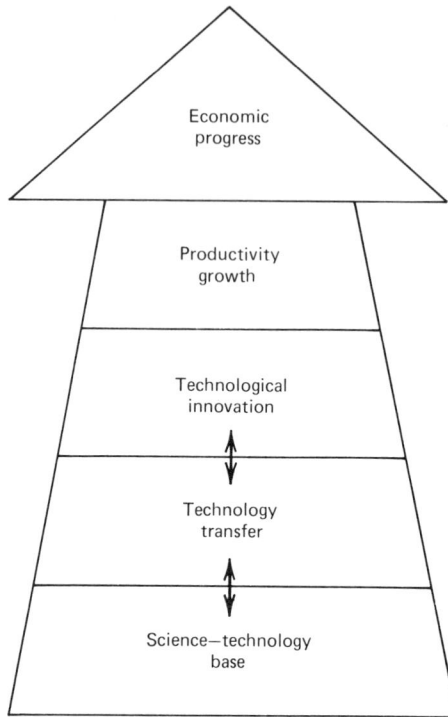

Figure 1-5   The technoeconomic interface.

personification of an innate ability to move easily within and between the realms of science, technology, and economics.[27] Nevertheless, he still needed to rely on a number of close associates to help him pick his way through complex technological, political, and financial hurdles. To place in proper perspective the principal elements of the technoeconomic interface, consider Figure 1-5. The science and technology base provides the principal foundation for achieving productivity growth and economic progress. However, the science and technology base, if it is to contribute in any significant way, must be effectively coupled to demands of the economic sector.

Technological innovation is the key link between economic progress and the science-technology base. Innovative products and services introduced into commerce help stimulate higher productive efficiencies that give rise to greater economic progress. A stronger economy in turn generates growing capital resources, a portion of which is plowed back for reinvestment in more innovative ventures and further productivity gains. However, the link between technological innovation and the science-technology base can not be taken for granted irrespective of the resources devoted to R&D. Full exploitation of the science-technology base depends on ready access to and the effective use of the scientific knowledge and technical know-how already in existence. There must be a conscious effort to utilize these resources through transfer and adaptation for improving the rate of technological innovation. Technological advances derived from the innovation process also contribute to an expanding science-technology base, which in turn provides a stronger foundation for further economic gains.

Of the four principal elements for economic growth illustrated in Figure 1-5, perhaps the least recognized and appreciated is technology transfer. A major reason is the feeling that we have successfully gotten by in the past without concerted efforts to transfer technology. This outlook is tantamount to expecting that future conditions will mirror those of the past. However, world conditions continue to change. New patterns of international technology flow will develop. Rapid dispersion of technical and management skills will take place, bringing on increased competition and shortened product life cycles. Demands for technological products and services will continue to expand in the face of strong competition for the R&D dollar. The information explosion will be exacerbated. All these factors will make the conscious transfer and effective use of available technology even more compelling and too important to leave to chance if strong rates of economic growth are to be restored and sustained.

Technology transfer acts to strengthen the linkage between the innovation process and the science-technology base. However, the flow of technology should not be construed to be only one way, that is, from the science-technology base. There is an equally important feedback link where technological progress resulting from the innovation process also acts to broaden the science-technology base in conjunction with the R&D input. In general, the balanced, bilateral flow of technology is an important aspect of technology transfer if substantive benefits are to be realized by the principal parties involved. Although the one-way flow of technology may produce transitory benefits, long-term advantages from technology transfer, whether at the industrial, national, or international level, are sustained only by means of a mutually beneficial exchange in which both the source and recipient prosper.

CHAPTER 2
International Aspects of American Technology

America began in the eighteenth century as a struggling, developing country dependent mainly on Europe for its technology. Improvisation, adaptation, and "Yankee ingenuity" marked the times. By the late 1700s, the industrial revolution in Europe was in full swing and spilled over onto the North American continent. The industrial revolution gave rise most notably to the beginnings of a shipbuilding industry in the New World and, by the early 1800s, to a small-arms industry that developed capabilites of mass production.[1] Industrialization spread rapidly to other industries, such as textiles and ironmaking. By the 1850s, American ironmaking handiwork competed with Great Britain in the world market despite the fact that the American economy was still agriculturally based and dependent mainly on imports for manufactured goods. By the late 1800s, industrial progress thrived as power-driven machines and factory assembly methods were more widely used to capitalize on production-line economies.

American entrepreneurs became noted for their abilities to harness technological assets for productive output. American ingenuity in applying technology to industrial practice was in many areas unexcelled in relation to European innovativeness.[1] American innovations, however, tended to be labor-saving because of the tight labor supply at the time, whereas European innovations tended to be resource conserving because of the limited availability of raw materials in Europe. Until World War II, trade between the United States and Europe centered on the use of cross-licensing agreements that effectively divided markets and limited competition. In technology-intensive industries in particular, where patents played an important role, cross-licensing between American and European firms acted to divide markets and formed the main avenue of international technology transfer at the time.

In the post World War II period the pattern of international technology transfer began to change as the United States began to export more manufactured goods to a war-devastated Europe. However, Europe's recovery was swift. By the mid 1950s, increased European competition emerged in the U.S. export markets. At the same time strong antitrust actions began to curtail the use of cross-licensing agreements widely employed at the time. Instead, direct investments abroad emerged as the principal means for technology flow to the extent that by the mid 1960s foreign direct investments were widely responsible for much of the technology transferred from the United States.

The postwar period also witnessed continued large U.S. spending on military programs and subsequently on space exploration. Significant civilian spin-offs from public investments in military and aerospace R&D began to appear in the commercial sector, for example numerically controlled machine tools, integrated circuits, commercial jet aircraft, and communications satellites. Commercial spin-offs such as these were aided by the sizeable government R&D spending that served to reduce the technical and market risks faced by U.S. industry. Technological innovations such as these also contributed to extending U.S. technological and economic inroads in Europe. Europeans were consequently prompted to respond to the "American challenge" by providing direct governmental support for commercial developments in major industrial areas such as aircraft, electronics, and nuclear power. By the 1960s, Great Britain, and later West Germany and France, poured huge sums into civilian developments in hopes of countering the U.S. influence. However, it soon became apparent that direct government support of commercial development projects was not financially sound as the rate of return from subsequent sales was inadequate to justify the original government investment. Large public expenditures on civilian development projects proved inadequate for stimulating the technical and economic climate needed for successful technological innovation. A major reason is that genuinely attractive commercial developments typically have little trouble attracting private capital, whereas only the marginal projects having higher technical and financial risks are generally proposed for direct government support.

The European experience suggests that the transfer of government technology to the private sector is perhaps a more effective channel by which government investments in R&D may be reflected in new commercial innovations. Consider, for example, government investments in military or aerospace R&D programs. As a minimum, progress in the form of improved national defense or advancing the frontiers of space-age science represents a justifiable return on expenditure of public funds; any further commercial spin-offs may be considered as an unexpected but welcome dividend. Considerable attention has been devoted to this aspect of technology transfer in the United States as will be detailed later.

The high levels of R&D investment in the military and aerospace sectors was a prime factor in the growth of technology-intensive industries in the United States. In addition, these industries are the leaders in exporting, licensing, and investing abroad. These avenues provide the technology-intensive industries the option of exploiting their technological resources by domestic production and export or by utilizing foreign-based production via licensing and direct investment. The choice is dictated mainly by economic factors, such as access to overseas markets, relative wage scales, transportation cost differentials, tariff restrictions, raw materials availability, and tax advantages. An additional not unimportant consideration is potential access to foreign sources of technology in the case of technology transfers to industrialized countries. Clearly, economic considerations have favored the establishment of foreign-based production judging from the rapid growth of U.S. multinational corporations in recent years.

Foundations of International Technology Transfer

Commercial transactions between American and foreign firms form the foundation for the international flow of technology. The principal motive for these transactions, however, is not technology transfer; rather, it is associated with economic incentives such as the prospect of making a reasonable return on investment or gaining market access for future investment opportunities. Nevertheless, much technology is transferred abroad as a matter of course because the business transactions typically involve setting up foreign production units and outlets for technologically based goods and services. They may involve foreign-based organizations in which the U.S. firm retains an equity interest, or they may involve totally unaffiliated overseas organizations. Technology transactions with overseas subsidiaries, however, are more likely to include the management and marketing know-how, and the necessary educational and training support needed to exploit the technology effectively.

A wide variety of avenues exist by which technology is transferred abroad. Generally, the mode of transfer depends on the particular technology involved as well as on the normal commercial considerations associated with the transaction. The efficiency of the transfer method, however, is heavily dependent on the level of communication and mutual understanding achieved between the seller and the buyer. The transfer of information is a necessary precondition for the effective transfer of technology. The need for effective communication consequently makes good interpersonal relations and person-to-person contact indispensable in successful cases of technology transfer. Also, as might be expected, the more complex the technology, the more active the communications and interpersonal relationships required to consummate the technology transaction.

Foreign direct investment and licensing are generally considered to be the principal channels by which technology is transferred overseas.[2] However, there are important distinctions between the two concepts that must be kept clearly in mind. Foreign direct investments are made primarily from commercial considerations such as the desire to locate a plant overseas in order to reduce distribution costs or to avoid import duties. Foreign direct investments provide an environment conducive to technology transfer. On the other hand, licensing represents a means by which technology is transferred. Licensing decisions are based primarily on the desire to buy or sell a specific technology or know-how. A license agreement may be concluded between two parties that may be affiliated or unaffiliated. In the former case, one party retains equity participation in the other party in whole or in part. In the latter case, the license agreement is entered into between two independent parties. When technology is transferred via a licensing agreement between two affiliated parties, both foreign direct investment and licensing channels are employed. Thus foreign direct investment and licensing are not mutually exclusive processes.

Foreign direct investment involves full or partial ownership in a foreign subsidiary by the parent firm. The parent organization typically provides capital, technology, management, and marketing skills while the foreign subsidiary provides material and labor resources. Foreign direct investments are effective channels for technology transfer because the equity interest in the foreign subsidiary by the parent firm constitutes a powerful economic incentive to insure that the transfer proceeds efficiently and effectively. However, from the standpoint of the recipient country, foreign direct investments are not very desirable because of the limited ability to learn and absorb the technology and management skills in the case where the parent U.S. firm retains complete ownership of the subsidiary. The desire for greater control over in-country enterprises of multinational companies has produced a growing trend towards shared-equity or joint ventures. These enterprises are partially owned by local public or private interests and partially owned by the parent multinational. Here, both parties contribute capital towards the new enterprise. The management, control, and profits are shared in proportion to each party's equity participation. Also, the degree of technological commitment by the parent organization to the new subsidiary is correspondingly reduced roughly in proportion to its share ownership. The trend in recent years towards increased foreign participations in U.S. investments abroad has consequently prompted American firms to emphasize more their management and marketing skills over design and production technologies.

The sale of turnkey plants abroad represents the extreme case where the parent U.S. firm retains no equity interest. In this case manufacturing plants are built and sold as a package to the foreign recipient nation including all the know-how and skills necessary for plant operation by indigenous workers. Turnkey plants therefore represent more correctly a contract sale rather than an equity investment on the part of the American firm. Turnkey plants are preferred by countries where only a limited technical and industrial infrastructure exists because the plants, when supported with the necessary training and technical assistance, afford a quick and efficient method for building manufacturing capabilities and a national industrial base. Of course, unless an agreement forbidding competition is made, the turnkey plant could eventually compete directly in markets served by the original firm that contracted to build the plant. Such was the case when Fiat built the Soviet Union's huge Togliattigrad auto plant and subsequently found that some of the Russian autos were then exported to Western Europe in direct competition with Fiat's own models.

The turnkey plant is basically a contractual sale by the party with the technical and management know-how and includes no ownership in the plant when completed. Nevertheless, the seller may be requested to provide management and technical assistance in the plant operations. Additional contractual arrangements such as management contracts and technical service agreements may be negotiated for these purposes. The management contract conveys planning and decision-making authority to the plant builder to manage and operate the plant for the life of the contract. The technical service agreement provides for additional technical assistance such as engineering studies, equipment installation, maintenance, and process improvement. Neither of these contracts involve equity participation by the contractor. Moreover, it is not necessary that the contractor engaged to provide management and technical services be the party that built the plant, although normal circumstances would dictate that the builder be retained for these services because of his firsthand familiarity with the plant design and construction. Management contracts and technical service agreements are effective mechanisms for technology transfer as they normally require a high level of personal interaction between the two contractual parties over an extended length of time. Considerable amounts of firsthand knowledge and experience in management skills, operations, and maintenance can be acquired by the local work force through these types of contractual arrangements.

For all the different international business arrangements possible, the actual mechanism by which technology is transferred is very often the licensing agreement. The license agreement is technology specific. It may cover a specific technology, patent, trade secret, know-how, manufacturing right, trademark, or sales and distribution rights for a particular product. Licensing conveys to the recipient or licensee the right to use the technology or item for a limited time and application. In return the licensor may receive a lump-sum payment, royalty payments in some fixed proportion of resultant sales, or both. The licensing agreement may also include provisions for extensive training programs. However, it involves no management control by the licensor. Still, the licensing agreement may include restrictions on the use of the technology item in question basically to protect the licensor's competitive position and markets. These restrictions may be in the form of limitations on the application or the reexport of the technology. From the perspective of the licensee, licensing technology is a desirable option because it provides access to key technologies at minimum cost. At the same time it helps to raise the licensee's level of technical competence when supported with an adequate training program. As a result, the licensee's competence may develop sufficiently rapidly to achieve a measure of competitiveness with the licensor on expiration of the license agreement.

Certainly, the export of technology-intensive products is the most common form by which technology is transferred overseas. In terms of dollar volume, the sale of technology-intensive products represents the dominant mode of technology transfer. However, it is not a very efficient method because the key technological content generally is difficult to extract and often constitutes but a small proportion of the overall product. Furthermore, detailed design know-how, manufacturing technologies, and trade secrets are generally unavailable even with the most astute reverse-engineering expertise. Even with detailed operational and maintenance manuals, developing insight into the design know-how and manufacturing techniques behind the product is generally a very difficult task. For these reasons, product purchase is the least desirable method for acquiring technology from the standpoint of the purchaser or recipient country. In contrast, sale of technology-intensive products is the preferred mode from the perspective of the technology source as manufactured products represent the highest value-added, and therefore the most profitable, embodiment of his technological capacity. It is not surprising then that product purchase to acquire technological secrets constitutes more a last resort measure in the absence of other options. Nevertheless, such situations are often encountered by manufacturers when attempting to obtain industrial intelligence relating to a competitor's product.

It is interesting to see how the United States compares with the other principal trading nations in Western Europe in terms of relative reliance on foreign-based versus domestic production to exploit technological resources. Looking at the aggregate record for the United States, United Kingdom, France, West Germany, Japan, and Italy, approximately two thirds of the combined receipts from license fees and royalties during the 1960s was accounted for by the United States. At the same time the United States accounted for only about one third of the combined six-country total for technology-intensive exports. This is indicative of the heavier outflow of technology per se from the United States in the form of foreign direct investments and licensing. The United States has also been more active in developing foreign-based production units during this time frame. The other five Western nations, on the other hand, have been more preoccupied with domestic production and the export of their output. A major reason for this difference is the divergence in wage scales in the United States compared to Japan and Western Europe. American manufacturers found it advantageous earlier to locate production units abroad because of the comparatively lower wage scales at the time. The higher wage scales in the United States could not be offset by higher productivity levels very easily. However, the more recent dollar devaluations coupled with the more rapidly rising wage levels in Western Europe and Japan have moderated the situation considerably in recent years. As a result, foreign investments in the United States have begun to rise rapidly.

Although commercial transactions centered around foreign direct investment, licensing, and manufactured exports constitute the principal channels by which the bulk of U.S. technology is transferred overseas, there are many additional, noncommercial avenues that also contribute to the technology flow from the United States. For example, government-to-government agreements between the United States and other countries are commonly employed to exchange technical information or to conduct cooperative R&D projects. The agreements cover a host of subjects such as energy, environmental protection, and national defense. They involve extensive contact between government representatives including frequent visits and meetings. Tours of industrial plants and government facilities by foreign specialists may be arranged, and domestic training and educational programs may be established. The information exchange, although supposedly mutually beneficial, nevertheless flows out from the United States in most cases by virtue of the U.S. leadership position in many areas of science and technology.

Participation in international meetings is another method by which scientific and technical information is exchanged with other countries. International meetings also serve as a convenient forum for developing contacts with colleagues in other countries, and they generally are an important means for establishing additional information sources in the participant's field of specialization. The United States is usually well represented in international meetings and conferences. While attendance at international meetings is subject to a number of factors such as scheduling, location, and the availability of travel funds, U.S. interest and participation in international meetings is expected to continue at a high level.

Scientific and technical publications provide another outlet for information exchange. The contributions by U.S. authors to the world technical literature has been substantial. From a large sampling of the world's leading scientific and technical journals, U.S. authors contributed almost 40 percent of the published material.[3,4] Furthermore, U.S. publications are widely referenced in publications of all nations. Citations to U.S. publications significantly exceeds the rate expected from the U.S. share of world publications. U.S. publications thus constitute a considerable outpouring of U.S. scientific and technical knowledge for world consumption.

Consulting firms are playing an ever more important role in transferring technology from the industrialized to developing countries. Their expertise is diagnostic in being able to provide to developing countries the evaluational and judgemental skills necessary in analyzing problems, determining technological needs, accessing the repositories of technology available in the industrial world, and providing the marketing and management skills that underpin all ventures. International consultants also serve as pathfinders for U.S.

firms in their efforts to penetrate virgin markets in foreign countries. Moreover, they are often the catalyst for major deals involving international transactions in technology; as such, they provide a stimulative channel by which technology is exported from the United States.

Patterns of International Technology Flow

A look at the U.S. record of international transactions involving direct investments, licensing, and export trade of manufactured goods provides a general picture of the international flow of technology across America's borders. In particular, the flow of royalties and fees between American and foreign firms reflect the value of technology transferred either through a direct investment or a licensing context. Furthermore, trade figures on manufactured goods, particularly technology-intensive products, also serve to register the flow of technology arising from U.S. international trade. These records, as will be readily evident, clearly testify to the preponderant net outflow of U.S. technology that has occurred since the 1960s.

One measure of the extent to which technology is transferred internationally is provided by the flow of royalties and fees between U.S. firms and foreign-based organizations arising from direct investments and licensing transactions. Royalties and fees are paid for the use of proprietary property such as patents, technical know-how, formulas, manufacturing rights, designs, copyrights, and trademarks. As such, U.S. receipts of royalties and fees are indicative of the amount of proprietary property exported. Conversely, U.S. payments of royalties and fees reflect the extent of foreign technologies imported into the United States. The net balance of royalties and fees then is a measure of the direction and extensiveness of the net flow of technology from direct investments and licensing.

Typically, U.S. receipts from royalties and fees has been considerably higher than corresponding payments.[3,4] The net balance (receipts minus payments) rose approximately 210 percent between 1966–1977, from about $1.38 billion to $4.28 billion. The sizeable receipts/payments imbalance is underscored when one observes that total U.S. royalty and fee payments for foreign technology amounted to only $140 million in 1966, rising to only $447 million by 1977. This data is therefore indicative of the heavy net outflow of American technology in recent years. Moreover, the major share of the U.S. technology outflow has gone to the industrialized nations of the world.

A closer look at the net balance of royalties and fees shows that foreign direct investments account for the major proportion of the technology outflow. Some 80 percent of the net balance is derived from foreign direct investment in 1966. By 1977 the foreign direct investment share rose slightly to 82 percent. Though the percentage rise is small, it represents a substantial rise in actual dollars by virtue of the sizeable increase in the net balance during the 1966–1977 period. These data therefore suggest that U.S. firms are meeting with greater success in seeking an equity interest in overseas transactions. Furthermore, the data may reflect the fact that entering into joint ventures may be achieving greater acceptance by foreign firms. This appears to be the case at least with Japanese firms. The direct-investment share of total U.S. receipts from Japanese firms rose from 38 percent in 1973 to 51 percent in 1977. Evidently, a more open Japanese policy regarding foreign investments has permitted increased inflows of foreign capital into Japan.

Whereas U.S. technology traditionally has gone mostly to unaffiliated Japanese firms until recent years, the opposite is true with regard to Canada. U.S. receipts from direct investments in Canada have always been at a high level. Of the total U.S. receipts from Canada, typically more than 90 percent are direct-investment related. However, a slowing of this trend is noticeable in recent years as the direct-investment share of U.S. receipts from Canada has fallen from 94 percent in 1975 to 92 percent in 1977. Apparently, Canada's recent effort to deemphasize the importation of foreign technology in favor of developing greater technological self-reliance is achieving results.

Based on the record of U.S. receipts from royalties and fees, the developed nations of the world are the major recipients of American technology exports. U.S. receipts from direct investments in Western European nations, Japan, Canada, New Zealand, Australia, and South Africa constitute the predominant share of total direct-investment related receipts. During the 1966–1977 period, U.S. receipts from direct investments in the developed nations rose from 73 percent to 80 percent of total direct-investment related receipts. Conversely, during the same period a diminishing proportion of total U.S. receipts from direct investment came from the developing countries. These trends indicate that demand for U.S. technology from the industrialized world is continuing unabated. Furthermore, the data suggest that growth in U.S. foreign direct investment during 1966–1977 has primarily come from the developed countries. Evidently, technology exports from U.S. multinational corporations (MNCs) continue to be directed primarily at the developed countries notwithstanding widespread publicity about MNC expansion into the developing countries.

Despite the growing trend of U.S. technology transfers abroad as reflected in the rising U.S. receipts surplus in royalties and fees from foreign-based organizations, there has also been a distinct trend during the 1970s towards more foreign investments in the United States. In 1973 and 1974 there were increases in foreign direct investments in the United States amounting to 20 percent per year, rising to a level of $21.7 billion in 1974.[5] Western Europe, Japan, and Canada accounted for over 85 percent of the 1974 total. Since 1974, the weakening of the U.S. dollar in relation to most Western European currencies as well as to the Japanese yen has tended to encourage even more direct investments by foreign firms. In addition, higher oil payments to OPEC members has produced increasing amounts of money being recirculated into the United States in the form of OPEC investments. Yet the increasing level of foreign direct investment in the United States has not generated any substantial increase in the U.S. payments of royalties and fees to foreign firms. If foreign direct investment is indeed a primary channel for technology transfer, a significant rise in U.S. payments in royalties and fees would be expected as a result of a higher rate of technology flow into the United States generated by more direct investments by foreign firms.

The explanation could be that foreign direct investments in the United States may be motivated more from a desire to acquire U.S. technology through outright purchase of high-technology U.S. firms rather than to transfer technology into the United States. However, the consensus drawn from four seminars conducted by the National Academy of Engineering on the impact of foreign investments on technology transfer in the U.S. pharmaceutical, electronics and computers, nonelectrical machinery, and petrochemical industries concluded otherwise.[5] Although technological considerations were important in decisions on whether to invest in the United States, the primary purpose for investment in the United States was to gain access to the sizeable American market. That is, foreign direct investments in the United States is driven more by market considerations than by the desire to acquire U.S. technology.[6] Furthermore, foreign investments in the United States are more likely to result in a net inward flow of superior foreign technology so that foreign firms can compete effectively in the American market. Typically, technology on balance flows from the parent organization to the subsidiary. However, market rather than technological factors appear to be of prime importance. These considerations therefore suggest that foreign direct investments can, but may not, be accompanied by the transfer of technology. However, when technology is transferred internationally, it often takes place between affiliated organizations in a direct-investment context.

Consider now the record of U.S. exports of manufactured products, particularly technology-intensive goods. The export sale of technology-intensive manufactured goods is perhaps the most common form of technology transfer. Technology may be embodied in the design or development of the product, the manufacturing process, or in the use of special materials. The sale of technology-intensive goods generally represents a higher dollar-volume transaction than is generated in royalties and fees from foreign direct investment and licensing. However, the technology embodied in the manufactured product is subservient to the product application, whereas technology is the principal comodity when royalties and fees are paid in licensing transactions. Hence the actual technology content in a product is low on a per-dollar-volume basis. It is also difficult to extract, particularly if the technology resides in the production process. Nevertheless, the technological content of the product is a major determinant of product competitiveness from both a price and performance standpoint.

The technological content of a product may be characterized in several ways. A common approach is to use R&D expenditures per sales dollar as a measure of technology intensity. A product belonging to a particular industry that devotes more than a certain percentage of its sales dollar to R&D may be classified as a technology-intensive product. The problem with this approach, however, is that it does not account for the distribution of R&D expenditures among different product areas within the industry. One way to avoid this difficulty is to determine the ratio R&D expenditure to sales dollar for each product category from all companies making the product regardless of industry classification. The Department of Commerce utilizes this product-category definition of technology intensity based on the U.S. Standard Industrial Classification (SIC) codes.[7] By determining the technology intensity for the SIC product classifications, the average technology intensity for all product categories is established. Products with technology intensity higher than the average are defined as technology intensive; those that fall below the average are classified as nontechnology intensive. The determination of technology intensity from product rather than industry data should therefore avoid undue distortions arising from highly diversified industries.

Table 2-1 describes the product classes and the corresponding technology (research) intensity ratios for the period 1968–1970. At the top of the technology-intensive category are missiles, spacecraft, aircraft, computers, and electrical equipment. The nontechnology-intensive products include items such as automobiles, farm machinery, and textiles. Clearly, products may migrate from technology intensive to nontechnology intensive, and vice versa, over a period of time. Furthermore, the distinction between technology intensive and nontechnology intensive is a relative one. That is, the threshold for dividing between high and low technology products may be set at a prescribed technolgy-intensity level other than the average for all products. Keeping in mind that different methods for characterizing the technology content of a product also affects the definition of technology-intensive products, it is readily evident that the concept of technology-intensive products is at best only a gross indicator of technology content of a product and that inferences and conclusions drawn as a result are subject to these qualifications of definition.

Using the Department of Commerce definition of technology intensity, the U.S. export record of manufactured goods for the period 1968–1977 is shown in Figure 2-1.[8] The export of both technology-intensive and nontechnology-intensive goods has been rising during the period, with the most dramatic increases occuring between 1972 and 1974 when the dollar was devalued. Since 1974 the rise in exports of both categories slowed appreciably as the full impact of the world recession and slowered economic growth in the main trading nations manifested itself. Throughout the period, technology-intensive goods comprised roughly 40 percent of total U.S. manufactured exports.[7] This figure compares with an aveiage of only about 28 percent for West Germany, Japan, France, and the United Kingdom and of only about 25 percent for the rest of the OECD countries. Thus the United States stands out as having an unusually high proportion of total manufactured exports being technology-intensive goods. Equally interesting is the fact that there is little difference in the technology-intensive share of total manufactured exports in the other OECD countries. That is, the relatively high levels of R&D investment relative to the GNP in countries such as West Germany, Japan, France, and the United Kingdom appear not to be reflected in any appreciably higher proportion of technology-intensive products in their manufactured exports in comparison with the remaining "research-poor" OECD countries.

In addition to the exceptionally high percentage of U.S. manufactured exports being technology-intensive relative to other OECD countries, U.S. technology-intensive goods also possess a higher technology intensity than

Table 2-1   Manufactured Product Classes by Technology Intensity,
1968-1970

Product Class	Technology Intensity Ratio (%)
Technology intensive	
Guided missiles and spacecraft	84.52
Aircraft and parts	12.41
Office, computing, and accounting machines	11.61
Electric transmission and distribution equipment; electrical industrial apparatus; communication equipment and electronic components	11.01
Optical and medical instruments; photos, watches	9.44
Drugs and medicines	6.94
Plastic materials and synthetics	5.62
Engines and turbines	4.76
Agricultural chemicals	4.63
Ordnance, except guided missiles	3.64
Professional, scientific, and measuring instruments	3.17
Industrial chemicals	2.78
Radio and TV receiving equipment	2.57
Nontechnology intensive	
Farm machinery and equipment	2.34
Motor vehicles and equipment	2.15
Other electrical equipment and supplies	
Construction, mining, and related machinery	1.90
Other chemicals	1.76
Fabricated metal products	1.48
Rubber and plastic products	1.20
Metalworking machinery and equipment	1.17
Other transportation equipment	1.14
Petroleum and coal products	1.11
Other nonelectrical machinery	1.06
Other manufactures	1.02
Stone, clay, and glass products	0.90
Nonferrous metals and products	0.52
Ferrous metals and products	0.42
Textile mill products	0.28
Food and kindred products	0.21
Total manufacturing	2.36

Source: Reference 7.

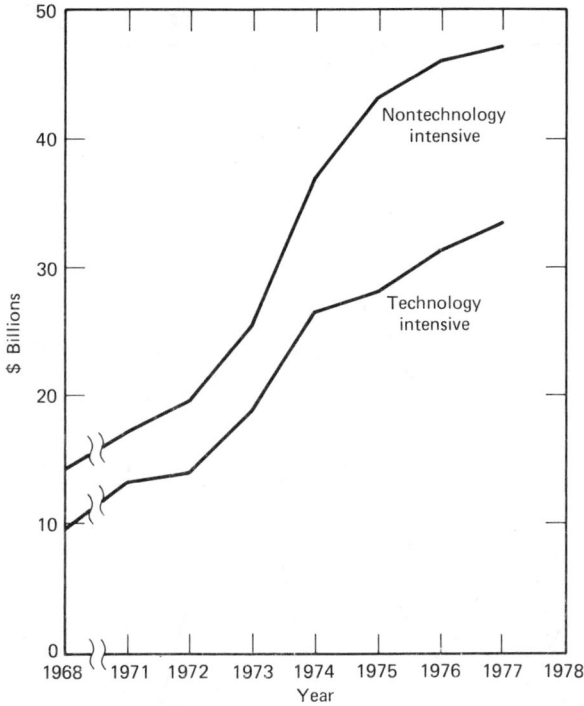

Figure 2-1   U.S. export of manufactured goods, 1968–1977. (Source: Reference 2-8)

in the other countries. In 1968 the average technology intensity of U.S. manufactured exports was 4.23 percent, compared with 2.36 percent for all domestic manufactured goods (Table 2-1).[7] In contrast, the technology intensity of manufactured exports from the other major industrial nations of the OECD was significantly lower: 3 percent in the United Kingdom, 2.7 percent in France and West Germany, and 2.36 in Japan. Thus technology-intensive exports from the United States represent a richer embodiment of R&D investment than from any other OECD country.

The U.S. trade balance (exports minus imports) in manufactured goods for both the technology-intensive and nontechnology-intensive sectors are shown in Figure 2-2 for the 1968–1977 period.[8] Despite the fact that technology-intensive goods make up less than half of total U.S. manufactured exports, they nevertheless yield a substantial trade surplus that by and large is sufficient to cover trade deficits in the nontechnology-intensive sector.

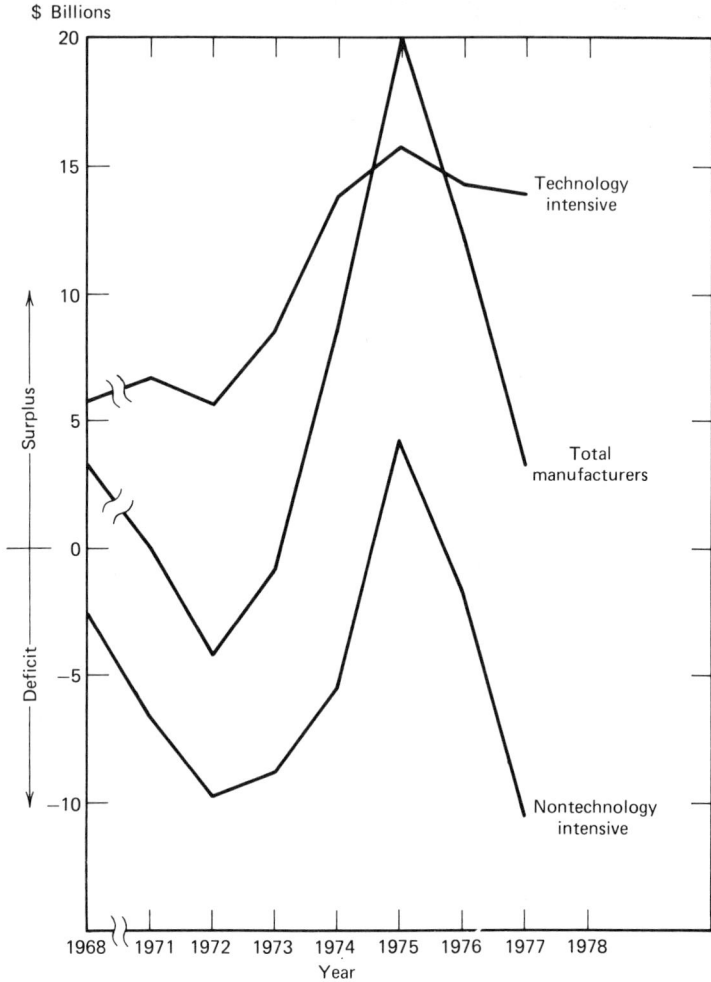

Figure 2-2*   *U.S. manufactured trade balance, 1968–1977. (Source: Reference 2-8)

United States trade is thus characterized as a net exporter of technology-intensive goods and a net importer of nontechnology-intensive products. The temporary suplus in nontechnology-intensive trade experienced in 1975 arose from a drop in imports experienced as a result of the recessionary economy in the United States. However, because of the differential recovery rates in the United States and abroad, imports climbed substantially in 1976 and 1977 to the extent that the trade balance in nontechnology-intensive products plunged to a record deficit by 1977. In contrast, the trade balance in technology-intensive goods held up fairly well during the same period, which is indicative of the lower price sensitivity and higher uniqueness factor of the technology-intensive sector compared with the nontechnology-intensive sector.

The unprecedented strength in the technology-intensive trade during the 1968–1977 period further belies the oft-stated association between R&D investment and U.S. trade in technological goods. Despite the decline of R&D investment in the United States starting in the mid 1960s and continuing into the 1970s, the trade balance in technology-intensive goods nevertheless compiled healthy surpluses, particularly since 1972, as a result of the dollar devaluations. This is perhaps not surprising in view of the considerable time normally required before research results emerge in the form of manufactured exports. The evidence thus lends further credence to the view that R&D investment is not a particularly useful determinant of the state of the national economy.

Looking at the trade balance for manufactured products on the whole, Figure 2-2 shows a wide-ranging fluctuation from the first postwar deficit approaching $5 billion in 1972 to a $20 billion surplus in 1975. The sharp swings in the trade balance in manufactured goods stem from the nontechnology-intensive trade, which is considerably more sensitive to price fluctuations and general economic conditions than the technology-intensive product group, which in fact serves as more of a stabilizing factor. The U.S. trade surplus in manufactured goods comes mainly from the developing countries. However, in 1977 this trade surplus with the developing countries was virtually wiped out by a correspondingly large trade deficit with Japan alone. In fact, Japan represents the singular exception to a generally favorable U.S. trade balance in manufactured goods.

The present trend towards more direct investment in the United States by foreign firms coupled with the already high level of foreign direct investment by U.S. firms illustrates the inexorably growing global interdependencies prevailing today. This global interdependency has been fueled in the main by the transfer of considerable technology from the United States over the last several decades. Not only will trade among the industrialized countries remain important to the United States, but the development of Third World countries will become increasingly intertwined with U.S. economic growth. Progress in the developing countries builds a more demanding and sophisticated export market for U.S. technology and goods, and it helps to generate a higher export consciousness on the part of domestic manufacturers. In addition, patterns of international trade and investment may be expected to shift as developing countries such as the OPEC members become more and more attentive to investment opportunities world-wide.[9] More direct investments by U.S. and foreign firms will result in greater dispersion of management, marketing and technical skills throughout the world. Also, the expanding role of multinational corporations on the international scene may be expected to accelerate further the pace of international trade in the technology-intensive industries.

U.S. Technology Export Implications and Issues

The persistent outflow of U.S. technology in recent decades stimulated in the main by commercial motives has nevertheless generated long-term implications and issues extending far beyond the more limited horizons of specific commercial ventures. Controversy has developed on whether the United States should continue its liberal technology trade practices of the past or promulgate new technology export policies to protect U.S. economic and national security interests. On the one hand, proponents of unconstrained commerce argue that technology exports are a natural consequence of the U.S. leadership position in science and technology and the application of science and technology to productive output. Furthermore, the export of technology stimulates international trade and helps the U.S. balance of payments. On the other hand, U.S. technology exports that help the formation of production facilities abroad displace U.S. exports of manufactured goods, eliminate jobs, increase trade competition, and contribute to a less favorable trade balance. Uncontrolled export of technology could also jeopardize national security by compromising technological advantage in advanced weapon systems. Consequently, issues relating to U.S. technology export policies have been drawing unusually intense scrutiny from high levels of both government and industry.

Debates on the subject generally distinguish between technology-intensive products and technology per se exported mainly as a result of foreign direct investments and licensing. This distinction appears important in attempts to analyze the implications of technology exports in both a civilian-commercial and military context.

The major issue from the civilian-commercial standpoint is whether the transfer of U.S. technology abroad helps or hinders economic welfare and domestic employment. Resolution of this issue, however, suffers from an incomplete knowledge of the relationships between the various modes of technology transfer and the factors of economic progress. The export of technological products has a much more salutary effect on the nation's balance of payments than technology transfer via foreign direct investment and licensing. In addition, competitive forces are more likely to be awakened in foreign countries as result of direct investment and licensing. Hence the export of goods represents the most preferred mode for international commercial transactions from the U.S. standpoint. However, changing factors of production such as relative wage scales, proximity to markets, and fiscal considerations have increasingly shifted the focus to foreign direct investment and licensing. Also, as will be discussed later, NATO geopolitical forces are working towards more U.S.-European shared ventures in the form of codevelopment and coproduction of major weapon systems. Unfortunately, present knowledge of the long-term effects of foreign direct investment and licensing on international trade and domestic employment is woefully inadequate. There is some evidence, however, that foreign direct investments tend to create greater demand for higher-skill jobs but may not be significant in generating more union jobs because union membership is concentrated in the nontechnology-intensive, import-competing industries.

The impact on the trade balance from technology transfers via foreign direct investments or licensing depends on whether the technology in question is supplementary to or competitive with domestic industries. Transfers that occur in noncompetitive areas benefit the U.S. economy in general as well as the exporting firm and the recipient country.[10] In this situation the transfer of technology not only produces foreign exchange earnings for the United States in the form of royalties and fees, but it also contributes to more efficient production units overseas which could mean lower prices of goods imported into the United States. However, transfers of technology that result in the establishment of production facilities abroad that are competitive with U.S. domestic production could have an adverse effect on the U.S. economy notwithstanding the economic benefits realized by the technology exporting firm. The major consequences of having competitive productive enterprises abroad are the displacement of U.S. exports in overseas markets and the introduction of competitive imports in the U.S. domestic market. The magnitude of these effects on the balance of payments, domestic employment, and general economic welfare tend to overshadow whatever short-term benefits might have been realized by the exporting firm from the original technology transaction.

There are some who consider the export of technology through foreign direct investment and licensing generally to hold long-term benefits for the United States. Yet the extent of these benefits are ill-defined, and the conditions under which foreign direct investment and licensing become disadvantageous still need to be determined. Therefore, in the present incomplete state of knowledge of the relationship between the transfer of U.S. technology abroad and its effect on the national economy, there is a natural reluctance to invite government intervention to discourage or impose restrictions on the transfer of U.S. technology overseas except in the interest of protecting national security.

From the military standpoint the control of U.S. technology exports is a necessity. The major issue is not whether to control but how best to control U.S. technology exports effectively to preserve the technological advantage affecting national security in areas where the United States enjoys technological superiority.

Technology export control has developed increased urgency in light of U.S.-USSR détente and new U.S. trade relationships with members of the Soviet Bloc and the People's Republic of China initiated in the early 1970s. Détente has permitted the Soviet Union to tap U.S. technology in many areas in which the Soviet Union lags, such as in computers, wide-body jets, agriculture, and deep-sea exploration. Valuable information has been acquired from studying detailed proposals submitted by competing U.S. firms seeking contracts to build sophisticated manufacturing plants in the Soviet Union. The sale of technology-intensive commercial products and the exchange programs under which Soviet scientists, engineers, technicians, and students come to the United States to study are additional mechanisms by which U.S. technology is acquired by the Soviet Union. In the U.S.-USSR student exchange programs, for instance, most Soviet students study engineering, computers, physics, and chemistry in the United States. On the other hand, American students generally study history, literature, language, and arts in the Soviet Union. In addition, the Soviet students have a wider range of colleges and universities to choose from in the United States than American students have in the Soviet Union.

East-West trade has gained new impetus as the United States and the communist world move from the confrontation and isolation of the 1950s and 1960s towards a normalization of relationships.[11] United States exports to the Soviet Union rose rapidly, beginning in the early 1970s, with the dollar volume more than tripling in value between 1971–1974 alone. Similarly, United States exports to the People's Republic of China has experienced even more dramatic growth, rising from virtually nothing in 1971 to about $1.2 billion in 1978. With the normalization of relations between the United States and the People's Republic of China in 1979, even higher levels of trade and U.S. exports may be expected. Looking at U.S. trade with the Eastern Bloc countries (Bulgaria, Czechoslovakia, German Democratic Republic, Hungary, Poland, Romania, USSR, and the People's Republic of China), a dramatic rise in trade activity has taken place since 1972 as illustrated by Figure 2-3. Clearly, the United States has been enjoying a net trade surplus vis-á-vis the communist countries. It should be kept in mind, however, that technology-intensive exports comprise a significant share of total U.S. exports to the East.

Concurrent with the rapid expansion of East-West trade, there has been an even more dramatic growth in "North-South" trade, that is, trade between the industrialized northern hemispheric nations and the developing nations concentrated in the southern hemisphere. In terms of trade in manufactured goods, the dollar volume of American exports to developing coun-

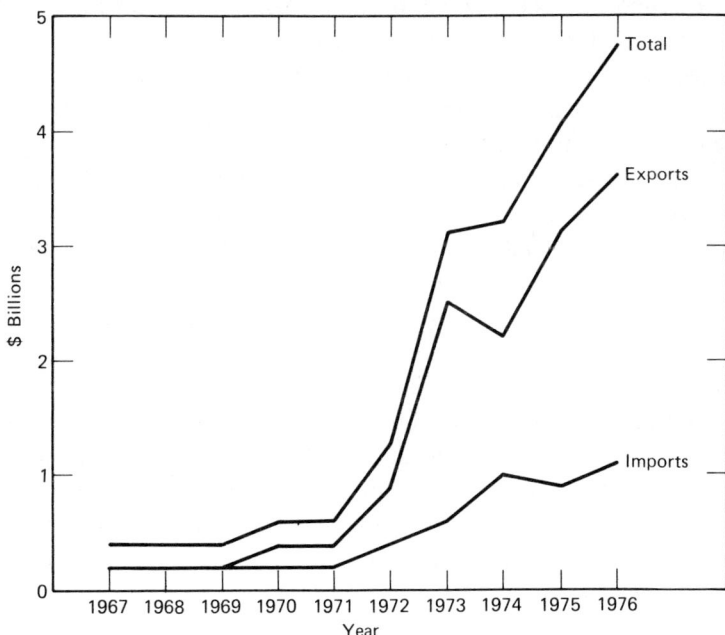

Figure 2-3   U.S. trade with communist countries (Bulgaria, Czechoslovakia, East Germany, Hungary, Poland, Romania, the Soviet Union, and the People's Republic of China). (Source: Reference 2-11)

tries in 1977 exceeded the combined total for Western Europe, Japan, and the communist countries.[12] Prospects are good for continuing rapid growth in North-South trade, in particular for trade involving the transfer of technology from the United States, as Third World countries seek to modernize and industrialize at a rapid pace. The growth of North-South trade further sharpens the need for effective technology export controls inasmuch as technology once exported to a friendly or neutral developing country could readily be reexported to a communist country in spite of all efforts to police against such retransfers. There have been numerous cases in which American arms and military equipment have found their way to unintended parties and countries through just such circuitous routes as third- and fourth-party countries.

The big attraction of course is American technology, particularly the design, production, and industrial management know-how. The unmistakable imprint that American technology has made on the industrial growth and development of the Western world since World War II has not been lost on the Eastern bloc. A similar transformation is being pursued by the centrally planned economies as considerations of technology has come rapidly to the fore in foreign affairs between East and West.

United States efforts to control exports date back to 1940, when controls were imposed to conserve domestic supplies needed for the war effort.[13,14] Subsequently, the Export Control Act of 1949 was enacted to deny exports to communist countries that could conceivably aid in their industrial development and war-making potential. American policy under this Act produced a virtual trade embargo with the Soviet Union. Because this policy did not substantively inhibit Soviet economic development and because the U.S. allies did not heed such a restrictive embargo, a policy of selective embargo was subsequently adopted by the United States in the form of the Export Administration Act of 1969. The Act applies to "dual-use" commercial items that may have military applications also. The Export Administration Act as amended establishes that the U.S. policy is to "encourage trade with all countries with whom we have diplomatic or trading relations" as well as "restrict the export of goods and technology which would make a significant contribution to the military potential of any other nation or nations which would prove detrimental to the national security of the United States."[15] Exports are authorized only after proper licenses are granted by the U.S. Government.

The Department of Commerce has principal responsibility for processing export license applications for dual-use items identified on the U.S. Commodity Control List (CCL). The State Department similarly processes export license applications for goods on the munitions control list in accordance with the International Traffic in Arms Regulation (ITAR). Cases requiring a closer assessment with respect to national security implications are forwarded to the Department of Defense, where they are further scrutinized from the standpoint of military relevance and potential impact on existing defense cooperation programs with U.S. allies. The daily case load is heavy. The Office of Export Administration in the Department of Commerce receives nearly 300 export license applications every working day; approximately 10 percent of these involve exports to communist countries.[14] Although the bulk of the license applications are processed by the Commerce and State Departments, approximately 6000 cases per year are nevertheless referred to the Defense Department for further review. The high case load, their complexity, and the need to coordinate actions among several agencies has meant in some cases inordinate delays in processing the license applications. As a result, considerable attention has been given to streamlining the license review procedures and shortening the processing time. Some progress has in fact been reported by the Defense Department in paring the average turnaround time for license cases from 29 days to 12 days.[15]

Control of technology exports to the communist world is a concern not only of the United States but of the U.S. allies. As a result, a multilateral control mechanism called the Coordinating Committee of the Consultative Group of Nations (COCOM) was formed in 1949, composed of all NATO member countries (except Iceland) and Japan. Its purpose is to establish among the member countries coordinative controls on exports of strategic products as identified by the COCOM list. The list is composed of 150 items that have both commercial and military value. COCOM activities principally involve negotiation of the content items of the list and the review of proposed transactions to be excepted from the embargo. The principal criteria to determine whether or not an item is considered strategic and therefore subject to controls are: (1) it has peacetime military use, (2) it contains extractable technology that is militarily significant, and (3) there exists a militarily significant deficiency in the Eastern bloc.[16] In reviewing a proposed exception from the COCOM list, the case is evaluated from the standpoint of potential diversion of the (1) product, or (2) embodied technology, to military application and (3) the likelihood that it would fulfill a critical deficiency in the East. Other judgmental factors, such as whether military use of the item in question has been made in Western countries, are also considered before arriving at each export control decision. COCOM control also extends to technical data although the most critical technologies have not yet been fully identified. Comparative to product end-items, however, the technical data case load is light primarily because manufacturers prefer to export products rather than technical data and also because the technical data tend to be associated with specific products. Although COCOM decisions are not legally binding on its members, mutual interests and concerns have made substantial agreement and cooperation possible. Nevertheless, in the minds of many, COCOM is more symbolic than substantive as cases of technology transfers to the communist world by COCOM members have increased markedly.[17]

The chief cause of less than effective COCOM controls appears to be differences in national laws and policies of the member nations. Whereas both technologies and products are subject to U.S. export controls, some COCOM members control only products and not technologies and technical know-how. United States export control law applies also to the reexport of U.S. technology by the recipient country. However, strict enforcement of this law is difficult despite prior assurances given that reexport controls will be observed. Furthermore, only U.S. firms are subject to prosecution in the event of violation. Another inconsistency is that the export control laws in some COCOM countries do not cover reexports to a third country. Therefore, strategic exports can find unimpeded transfer to communist countries through this route. Differences in national interests also promote differing interpretations of COCOM list restrictions. As a result, items which are considered embargoed from communist countries by one COCOM member may not be so interpreted by a second COCOM member. It must also be recognized that not all countries of the industrialized West are members of COCOM. Active trading nations such as Switzerland and Sweden, for example, are not members, nor are most developing nations included in the CO-COM agreement. It is not difficult to comprehend then the difficulty of the task and the inadequacies of the COCOM agreement. Yet it is the only multilateral mechanism available to the United States and its allies for establishing cooperative controls on technology exports.

To strengthen the U.S. export control process, the Department of Defense in the spring of 1974 asked the Defense Science Board to form a task force to look into currect U.S. technology export control policies and practices and to make appropriate recommendations.[18] The task force was constituted with representatives from government and industry and was organized into four subcommittees responsible for the following high-technology areas: airframes, aircraft jet engines, instrumentation and solid state devices. The principal recommendation of the task force was that effective control of design and manufacturing know-how was vital to maintaining U.S. technological superiority because they impact on the nation's strategic technological capabilities more critically than do other forms of technological endeavor. Furthermore, the task force recommended controlling the export of technology per se as distinguished from products of technology except for those critical products having direct military significance. Technologies are described as either revolutionary or evolutionary in nature. Revolutionary advances should be subjected to close scrutiny and generally should not be exported in order to preserve U.S. strategic lead time in that technology area. Technologies that represent evolutionary advances, on the other hand, may be approved for export depending on the specific circumstances of each individual case, such as the foreign country involved, their export policies, and their relative position with the United States in the technology area in question.

The Defense Science Board task force also recommended that the United States needs to define its export control policy objectives, develop a simpler criterion for making export control decisions, and should adopt a more pragmatic approach to the license-application review process. The review process for U.S. export license applications should emphasize controls on technology and know-how rather than products. It should also take into account the mode of technology transfer, giving more scrutiny to the more active modes, such as the sale of turnkey plants or joint ventures as opposed to more "passive" modes of technology transfer, such as product sales or distribution of commercial literature, inasmuch as the more active modes are usually also the more effective ones. Likewise, the United States should adopt a more conservative stance with respect to the COCOM agreement in view of the difficulty in policing the eventual disposition of exports to CO-COM countries. The COCOM list also needs to be pared to only technological products of direct military significance so that adequate attention and resources may be devoted to controlling the more significant items. Furthermore, the United States should not depend on the COCOM agreement for absolute control of critical technologies beyond its borders.

The findings and recommendations of the Defense Science Board task force have provided the foundation for subsequent efforts to tighten U.S. technology export controls. The Defense Department in 1977 issued an interim policy statement indicating that the main objective of export controls is "to protect the United States" lead time relative to its adversaries in the application of technology to military capabilities." The distinction between technology-based products and the technology per se was adopted, with the major emphasis placed on controlling exports of design, manufacturing, operational, and maintenance know-how as well as test and inspection equipment. Although the controls applied to technology exports to any country, specific exceptions are made in the case of critical technology exports to allied countries in order to strengthen collective security and to promote NATO weapons standardization and interoperability. The revised policy loosens restrictions on the export of certain end products but tightens controls on the export of technologies that can be used to upgrade military capabilities. Revisions will also be made in the COCOM list to reflect the current policy. Government task groups have already been formed by the Defense Department for this task.

Industry groups have also been formed to assist the Defense Department in establishing and implementing a clear set of criteria and guidelines for identifying critical technologies of significant military value on which U.S. military superiority is considered most dependent. Initially, 15 critical technology areas have been identified: computer network, large computer system, software, automated real-time control, composite and defense materials processing and manufacturing, directed energy, LSI-VLSI* design and manufacturing, military instrumentation, telecommunications, guidance and control, microwave component, military vehicular engine, fiber optics and advanced optics, sensor, and undersea system.[15] The 15 critical technology areas are to be further broken down into the subset of key technologies, component equipment, and technical data deemed critical to national security. This listing of critical technologies, products, and technical data will presumably be short and manageable and easily subjected to stringent export controls. However, establishment of the criteria for criticality and the identification of the critical technologies, products, and data has proven to be complex and difficult. Furthermore, it is not yet clear just how the list of critical items would be integrated into the present export control process and the U.S. commodity control list or the COCOM list.

The keystone of the Defense Department's export control policy is the protection of U.S. lead time relative to adversary countries for technologies that are militarily significant. Of major importance is the assessment of potential contributions that a technology could make in advancing adversary capabilities. Assessments are complicated by the time-dependent value

---

*LSI stands for large-scale integration and VLSI for very large scale integration in reference to microelectronic circuits.

of technology. That is, there generally is a definite life cycle associated with new technologies and technological end products. The extent of the life cycle is dependent on the obsolescent rate for the technology in question or the rate at which competitive products are introduced. It depends on the rate of development of emerging technologies or products that would eventually supplant the technology or compete with the product. United States lead time in critical technologies also depends on the relative ease with which adversary countries could acquire comparable technologies from other countries. Lead time estimates therefore require a comparative assessment of geopolitical as well as military and technological factors in the United States vis-à-vis other countries. Furthermore, a basic understanding of military missions and doctrine and the potential contributions of specific technologies to mission needs must be established to adequately assess criticality. Of course, criticality judgments are also a function of the individual biases and perceptions of the person making the assessment. For these reasons, it is easy to appreciate the considerable difficulty encountered in formulating suitably general criteria and guidelines for establishing criticality without requiring a case-by-case analysis.

In summary, United States experience in international technology transfer has been primarily that of a supplier of technology to the rest of the world. The mainstay of U.S. foreign trade has been the export of technology-intensive products from which the large part of U.S. foreign trade earnings are derived. The profound and unabated outflow of U.S. technology has given rise to a gamut of complex issues relating to national security, foreign policy, international trade, foreign competition, domestic employment, and market access. Against this backdrop has arisen a divergence of views concerning the direction of future U.S. technology export policies. An in depth reexamination of present policies has been initiated to attempt to determine the long-term implications of U.S. technology exports on the domestic economy, national security, and international relationships so that public policy options may be formulated on a sound basis. The tenor of these activities has been concerned with developing more effective controls on U.S. technology exports. Implicit in this stance is that the United States should maintain its past technology export orientation albeit from a more judicious and astute policy framework. However, this approach overlooks an important option that, in essence, can turn the consequences of past U.S. technology outflows into a decided advantage—namely, to develop a balanced return flow of foreign technology. Clearly, in those areas where the United States is ahead, the U.S. technological advantage should be preserved by slowing the transfer abroad of U.S. leading-edge technologies. However, in those technology areas where the United States is behind, the primary effort should be to try to catch up by learning and adopting wherever possible the fruits of foreign technological advances. This option must not be neglected as the United States struggles to reinvigorate its economy and reestablish its former pre-eminence in technology-based innovations. Before considering this option in greater depth, however, it is advisable to establish a better understanding of the consequences of past out flows of U.S. technology abroad in terms of developing increased technological competence and competitiveness in foreign countries.

CHAPTER 3

Consequences of
American Technology Exports

The enormous outflow of American technology since World War II has to a large extent shaped the global relationships and political, economic, and military challenges facing the United States today. American technology has been the keystone by which other countries have modernized and industrialized. Developed and developing countries alike have made sizeable economic strides as a result of American technology and know-how. Foreign adaptations and improvements on U.S. technology furthermore have produced stout challenges to U.S. technological leadership and keener foreign competition in commercial areas such as consumer electronics and civil aircraft.

Nowhere is the slippage in the American technological edge more apparent than from the growing competitiveness to U.S. manufactured goods evidenced by the rising tide of foreign manufactured imports on the domestic scene. Home video tape recorders marketed in the United States are made in Japan despite the fact that the United States pioneered its development. From Japan also come television sets, tape recorders, cassettes, automobiles, electron microscopes, steel, and cameras. Where roughly a third of the television sets sold in the United States in 1970 were imported, by 1979 foreign imports accounted for more than half of the domestic market. In the nuclear power industry, the United States is no longer dominant. Sweden, France, Canada, and the Federal Republic of Germany must also be considered. Likewise in chemicals, American companies are encountering stiff competition from German firms. Most notably, in the aircraft industry where U.S. manufacturers previously competed comfortably in world markets, America has abdicated the supersonic transport market to the British, the French, and the Russians. Additionally, more conventional aircraft of foreign manu-

57

facture, such as the European airbus, are grabbing a growing share of the world commercial aircraft market. In 1978 alone, Airbus Industrie, a French, West German, British, Dutch, Belgian, and Spanish consortium, managed to more than triple their share of world orders from 6 to 19 percent. This gain was made at the expense of the major American commercial aircraft manufacturers, Boeing, Lockheed, and McDonnell Douglas, which experienced declines in market shares.

Competition for U.S. manufactured goods comes not only from industrialized countries but also from the developing countries. The volume of U.S. exports to the non-OPEC developing countries generally exceed exports to the European community. However, where there used to be comfortable trade surpluses, imports have risen to the point that negative trade balances were incurred in 1977 and 1978. The imports furthermore were not made up only of raw commodities, but included manufactured goods—such as shoes, textiles, steel, and radios—that accounted for roughly half of U.S. imports from the non-oil-producing developing countries. This rise in manufacturing capability comes from intense efforts of the developing countries to try to capitalize on their comparatively low-wage scales and availability of raw materials to achieve greater industrial progress and fuller utilization of the indigenous labor base. Labor-intensive manufacturing plants are increasingly being shifted to the developing countries as a result. In particular, South Korea, Taiwan, Singapore and Hong-Kong have shown rapid progress in attracting manufacturing industries. These countries along with Brazil and Mexico, have in fact managed to raise their collective share of the world manufacturing exports from 1.5 percent in 1963 to 5 percent in 1976. As industrial development continues to be vigorously pursued by the developing countries, competition to American manufactured exports in the labor-intensive industries can be expected to intensify further.

Increased foreign competitiveness to American goods in the domestic and global markets has been the most visible outgrowth of the history of high U.S. technology exports. As the economies of Western Europe, Japan, and the United States reach comparable maturity, they become more dependent on each others' markets. The increased foreign competition stems not only from their generally rising level of technological competence but also from their demonstrated ability to innovate and to translate their technological know-how into competitive products and services. The effect of stronger foreign trade competition has been magnified by the concomitant slackening of U.S. productivity growth, which has adversely affected U.S. trade competitiveness even further. Although the policies and approaches employed in the different countries for the development and application of technology and the stimulation of innovation are generally dissimilar, the important yardstick from the U.S. standpoint is nevertheless their demonstrated success in making appreciable inroads into heretofore U.S. dominated markets.

Consequences of past U.S. technology export practices are by no means limited to the economic sphere. Equally important and far-reaching effects have been felt in the political and military arenas. The NATO (North Atlantic Treaty Organization) alliance in particular is undertaking extensive efforts to improve technological and economic cooperation between member nations for the common security. The principal objective concerns utilizing alliance resources in the most efficient manner to upgrade combat effectiveness of NATO military forces. Issues of rationalization, standardization, and interoperability (RSI) of conventional armaments and equipments have come to the fore as European members of NATO seek to increase their share of the development and production of armaments and equipment for NATO. The development, production, and deployment of competitive weapon systems and equipment that are not standardized or interoperable impair the combat effectiveness of the NATO operational forces and contributes to inefficient utilization of available alliance resources. Whereas the United States was the principal supplier of armaments and equipment to NATO in the 1950s and 1960s, the 1970s have witnessed the emergence of a strong European defense industry capable of producing a broad array of weapon systems with an established sales base in non-NATO countries.[1] The Europeans view their technological strengths to be second to none in areas such as tactical missiles and aircraft design. The development of the European defense industry has led to demands for greater cooperation and participation in the development and production of NATO conventional arms and equipment. As a result, a two-way transatlantic street in NATO defense equipment is being sought to capitalize effectively on the technological and industrial strengths on both sides of the Atlantic.

The net outflow of U.S. technology of the recent past has therefore precipitated far-reaching effects on the world economic, political, and military fronts. A closer examination of specific indicators and their international comparison would be helpful in gaining a better insight into the pervasive effect U.S. technology transfers abroad have had in shaping the world technological and economic environment.

Rising Foreign Technological Activity

Stiffer trade competition in technology-intensive goods is the most tangible evidence of growing technological activity in the foreign countries. Other indicators exist that point to increased foreign technological activity. One such indicator is the record of patents filed in foreign countries by U.S. residents compared with patents filed in the United States by foreign residents. Patents are an indication of a country's inventiveness and, as such, relate to the country's ability to introduce innovative goods and services to the commercial market. Because of the high costs associated with filing and maintaining patents in foreign countries, patents filed in more than one country are generally regarded to be more significant from the standpoint of possessing attractive market potential. Patent protection is primarily sought in those countries that the inventor perceives to constitute significant potential markets for his invention. Consequently, foreign inventors filing for U.S. patents are seeking to protect their technological advantage in the U.S. market. Another reason for seeking patent protection in other countries is that it may make the patent more attractive to potential licensees or buyers if the invention is protected in more than one country. Filing foreign patents helps protect the domestic and foreign markets from incursions from unlicensed, pirate manufacturing operations that may be set up in other countries.

The share of U.S. patents filed by foreign residents has been in a steady uptrend since the mid 1960s. As shown in Figure 3-1, the percentage of U.S. patents awarded to foreign residents has grown from 18 percent in 1963 to over 36 percent in 1977.[2,3] Japan and West Germany account for almost half of the total U.S. patents of foreign origin, while the United Kingdom and France account for another 20–25 percent. Hence almost 75 percent of all U.S. patents of foreign origin go to residents of these four industrialized countries. Japan has exhibited the most rapid growth, rising from 1 percent of total U.S. patents in 1963 to 9.5 percent in 1977. Furthermore, the rise in foreign patent activity in the United States extends over a wide range of industrial products in the electrical, chemical, and mechanical fields.

Figure 3-1 then is evidence of the growing technological competence in foreign countries and the conviction of these countries that a significant market for their technology-based products exists in the United States. At the same time there is evidence that the level of U.S. patent activity is slipping. In 1977, fewer patents were granted to Americans than anytime during the past 15 years. Moreover, in 1965, approximately 37 percent of all patent applications filed internationally were of U.S. origin. However, by 1974, the United States accounted for only about 29 percent of the total international patent applications. This slippage in all likelihood does not reflect a diminished global market assessment by American innovators so much as it is symptomatic of the growing foreign technical competence and America's declining ability to compete in international markets.

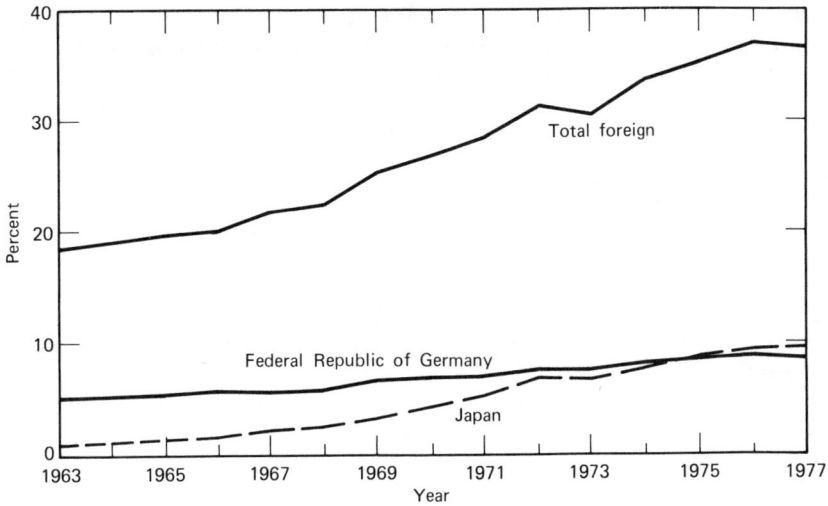

Figure 3-1   U.S. patents granted to foreign inventors. (Source: References 3-2 and 3-3)

Indeed, there appears to be some relationship between a nation's patent activity and its international competitiveness. As shown in Table 3-1, a look at the international patent activity of the major OECD nations compared with their respective manufactured exports activity during the period 1965/6–1975 shows a definite correlation.[4] The two columns on the left indicate the share of patent applications filed by country as a percentage of the total world patent applications for the years 1965 and 1975. The two columns on the right indicate the respective countries' share of total OECD manufactured exports for the years 1966 and 1975. As is readily evident, a declining trend for manufactured exports prevails in the United States and the United Kingdom, where the international patent activity is also in a downward trend. Conversely, in the remaining countries shown, a rising trend in manufactured exports accompanies a similar trend in their international patent activity. This evidence may be considered tentative at best. It could also be argued that the OECD export statistics shown do not wholly represent the international competitive stance of the countries shown. Still, the indicated correlation between international patent activity and manufactured export data appear to be more than coincidental.

Patent activity itself should be viewed as only an approximate measure of technological activity for several reasons. Although considerable technology

Table 3-1   Trends in International Patent Applications and
Manufactured Exports for Selected OECD Countries, '1965/6-1975

Country	Share of World Patent Applications (%)		Share of OECD Manufactured Exports (%)	
	1965	1975	1966	1975
France	7.2	7.4	8.6	10.2
West Germany	18.3	19.2	19.3	20.3
Italy	2.5	3.2	6.9	7.5
Japan	3.0	8.7	9.7	13.6
United Kingdom	11.7	7.7	13.4	9.3
United States	36.5	29.3	20.1	17.7

Source: Reference 3-4.

is disclosed only in patents,* nontechnological inventions, or those containing low technological content, are as likely to be patented as inventions resulting from technological activity. In addition, many significant technological advances are not patented because the invention represents only a marginal advance in the prior art, the inventor desires to maintain trade secrecy or avoid patent filing costs, or the technological advance is considered inappropriate for patenting such as generally encountered in computer software development. In some technical areas, such as in electronics where the obsolescent rate is rapid and patented products can often be "designed around" or by-passed with little difficulty, avoiding the patent system and protecting his invention as a trade secret may be the only available recourse for the inventor.

Underlying the growing foreign technological competence is necessarily a strong commitment by foreign countries to promote R&D and to strengthen their respective internal technological infrastructures. A look at the ratio of R&D expenditures as a percentage of gross national product in the industrialized countries confirms the stronger R&D commitment in foreign countries compared to the United States. Whereas the United States has typically been the leading country in terms of devoting the largest share of its economic output for the performance of research and development, the lead relative to the other industrialized nations has been considerably narrowed

*The U.S. Patent and Trademark Office estimates that 8 out of 10 patents contain technology not published in the nonpatent literature.

since 1964, principally on the strength of the precipitous decline in the U.S. R&D/GNP ratio in recent years. Figure 3-2 illustrates the national R&D/ GNP trends in the leading Western industrialized nations.[3] West Germany and Japan show decidedly strong upward trends, whereas France and the United Kingdom experienced a decline in their R&D/GNP ratio over the same period, but at an appreciably slower rate than that of the United States. Other Western nations have generally managed to increase or at worst maintain a stable level of R&D investment relative to their respective GNP. Although the trends in the R&D/GNP ratio show that other countries are narrowing the gap with the United States, it should be kept in mind that the absolute level of R&D funding in the United States is still at a high level compared to other countries.

Certainly, R&D investment in itself is but an input factor into the process for technological innovation and is not necessarily indicative of technological sophistication. R&D investment does not determine the relative efficiencies of different countries for translating technical know-how into commer-

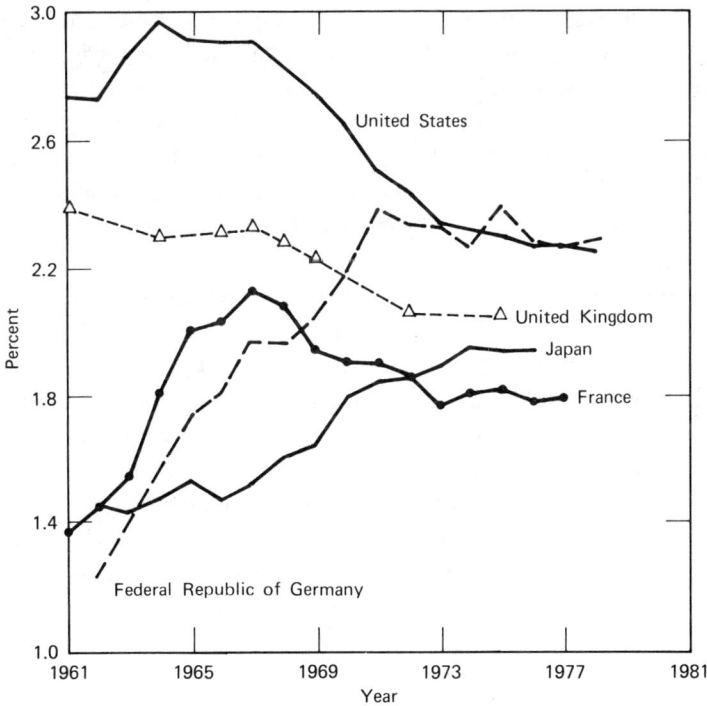

Figure 3-2   R&D expenditures as percent of GNP. (Source: Reference 3-3)

cial goods and services. It mirrors only the resources committed by various countries to promote technological progress. However, the more favorable R&D trends in foreign countries compared to the United States are significant because they occur at a time when the United States is exporting much of its technology. The implication is clear that the relatively stronger foreign commitment in R&D has provided the wherewithal by which American technology is adapted and upgraded into improved commercially competitive products. The resultant effect is that the foreign countries' technological prowess is enhanced and their national systems for technological innovation become more efficient.

Another indicator of a nation's commitment to the R&D effort is the extent of professional manpower devoted to science and engineering. Among the leading industrial nations—Canada, France, West Germany, Japan, the United Kingdom, United States, and the Soviet Union—the United States is notable in being the only country experiencing no growth in the number of scientists and engineers engaged in R&D per 10000 population during the period 1965–1973, primarily because of a decline in the professional manpower in the defense and aerospace sectors.[3] Since 1973, the R&D manpower picture in the United States has stabilized and has even shown a slight rise in 1978. In the other six countries, the numbers of scientists and engineers in R&D grew faster than the national population, with the Soviet Union exhibiting the highest rate of growth. Though by no means conclusive, the professional manpower picture nevertheless points to a rising R&D commitment in foreign industrialized countries compared to the United States.

We must recognize that funds and manpower devoted to R&D investment are necessary but not sufficient conditions for improved economic progress and international competitiveness. They represent resource inputs into national technological infrastructures and as such are subject to different stimuli and inefficiencies inherent in different national systems. They are indicative of the level of national effort devoted to technological activities. However, the output of each country's technological efforts is also dependent on widely divergent factors that influence the relative efficiencies of different technoeconomic systems. That is, the level of R&D investment will contribute to expanding the science-technology base, but depending on national capacities for technological innovation, the R&D effort may or may not be effectively reflected in improved economic performance. To the extent that a reasonably efficient technoeconomic system exists, R&D input factors are indicative of the technological prowess of a nation.

Patent activity is also an incomplete descriptor of a nation's ability to innovate. Although patent activity is a measure of inventive output, it nevertheless represents only a subset of the total technological activity of a nation. Furthermore, inventions in themselves are typically an input factor in the technological innovation process and not necessarily representative of the national capacity to translate technological assets into productive output. Therefore, to develop better insight into the abilities of different countries to capitalize effectively on technological resources for productive output, attention must be focused on the output factors of the technological innovation process.

Greater Foreign Innovativeness

The predominant export of American technology coupled with the stronger trends in R&D investment in other industrial countries provide foreign countries with the basic ingredients for substantially enhancing their respective abilities to innovate. The opportunity exists for foreign countries to employ to advantage the fruits of Yankee ingenuity and know-how and at the same time to develop a more judicious and complementary application of their own R&D resources in order to spur technological innovation. A determination of their comparative success in this endeavor would necessarily call for an examination of the outputs of their technoeconomic system— in particular, the comparative facility of different countries to produce technological innovations. Despite the substantial differences in national objectives, policies, institutions, and socioeconomic outlook in different countries, a convenient and useful assessment of the comparative ease with which technological innovations are produced can be made by examining the average innovation period in different countries.

The innovation period is the time that it takes for an innovation to move from first conception to market introduction. As such it provides a convenient measure of the relative innovative performance of different countries (as well as industries and organizations) when averaged over a sufficiently large number of innovations. The reason is that the confluence of all the technological and nontechnological factors that act to stimulate or impede the innovation process will be reflected ultimately in the innovation period. Innovations that meet with a minimum of impediments will naturally take less time than those that encounter substantially greater difficulties. Of course, the innovation period also depends on the specific nature of the innovation. For this reason, the innovation period when averaged over an appropriate field of innovations should provide an effective measure of the relative abilities of different countries to produce innovations independent of the nature of the innovation itself. That is, the average innovation period is indicative of the effectiveness of different national technoeconomic systems for directing national technological assets towards achieving greater economic progress. Countries exhibiting, on the average, a shorter innovation period are generally more proficient in the art than those requiring a longer time. A shorter average innovation time means a higher rate for technological innovation and an ability to respond more quickly to changes in market conditions, making possible improved economic performance in terms of higher productivity growth and greater market competitiveness.

A comparison of the average innovation period for the United States, United Kingdom, Japan, and the Federal Republic of Germany (FRG) can be made based on data accumulated on the 500 significant innovations of the 1953–1973 period described earlier (Chapter 1). Table 3-2 gives the number of innovations and the average innovative period calculated for each country.[5] Only those innovations which depended on a single technology source were considered because of the paucity of data on foreign innovations that utilized multiple sources of technology. Because there were only 11 innovations each of Japanese and German origin that depended on a single source of technology among the 500 cases considered, the confidence factor associated with their respective average innovation period is not as high as that for the United States and the United Kingdom. Still, a first-order comparison of average innovative periods can be made. As shown in the table, Japan and Germany exhibit a substantially shorter average innovation period compared to the United States and the United Kingdom,

Table 3–2   Comparison of Average Innovation Period for Selected Countries

	United States	United Kingdom	Japan	Federal Republic of Germany
No. of Innovations	65	36	11	11
Average Innovation Period (years)	7.4	7.7	3.4	5.2

Source: Reference 3-5.

which suggests that a more stimulative environment for technological innovation prevails in Japan and Germany. When considered in light of the acknowledged high competitiveness of Japanese and German products in the international market today, these results tend to substantiate the usefulness of considering the average innovation period as a measure of the relative innovativeness of different nations.

It is also interesting to note that the average innovative time period of the four countries shows an inverse correlation with the number of Nobel prizes awarded. That is, the country with the highest number of Nobel prize laureates displayed the longest average innovation period and vice versa. This observation seems not only to substantiate the notion that no close relationship exists between basic research results and innovational output, but also to indicate that a research orientation is antithetical to good innovational performance. More likely, the data suggests that, although basic research results may be an input into the innovation process, many other more important factors also affect the process and have a stronger bearing on the outcome.

The substantially shorter average innovation time in Japan and Germany provides these two nations with a distinct competitive advantage over the United States and the United Kingdom inasmuch as the shorter time span means more timely responsiveness to market demands. The data compiled on each of the 500 significant innovations were not sufficiently detailed to enable identification of the main reasons for the shorter innovation times found in Japan and Germany. Yet it is possible to examine in more detail the extent to which different countries depended on domestic versus foreign inventions, and from this perhaps some tentative conclusions can be drawn concerning the relative importance of international technology transfer in the different national technoeconomic systems.

Table 3-3 summarizes data on the origination of the inventions underlying the 500 significant innovations. The number of innovations per country is shown on the left-hand column in parentheses. The numbers given in each row across the table are the percentages of each country's innovations dependent on inventions derived from the different countries shown at the top. For example, 4.7 percent of the 85 innovations in the United Kingdom came from U.S. inventions. The last column on the right sums the percentages on each row representative of foreign-derived inventions (i.e., exclusive of the main diagonal). Looking at the U.S. innovations on the first row, we see that only 6.5 percent of U.S. innovations originated from foreign inventions.

Table 3-3 Percentage of Innovations Derived from Domestic and Foreign Inventions

Innovations by Country (No.)	Source of Inventions							
	United States	United Kingdom	Federal Republic of Germany	France	Japan	Canada	Other	Foreign Dependence
United States (319)	93.4	2.2	0.9	0.9	0.3	0.3	1.9	6.5
United Kingdom (85)	4.7	88.2	2.4	1.2	1.2	0	2.4	11.9
Federal Republic of Germany (33)	9.1	0	78.8	3.0	0	0	9.1	21.2
France (21)	0	0	0	100	3.0	0	0	0
Japan (34)	8.8	0	5.9	0	85.3	0	0	14.7
Canada (8)	12.5	0	25	0	0	62.5	0	37.5
Impact on foreign innovations	35.1	2.2	34.2	5.1	1.5	0.3	13.4	

Next to France, where all their 21 innovations depended on domestic inventions, the United States was the least dependent on foreign invention technology. On the other hand, Canada was the most dependent on foreign inventions (37.5 percent), followed by Germany (21.2 percent) and Japan (14.7 percent). Canada's relatively high dependence on foreign invention technology may appear surprising at first, but it is corroborated by independent data.[6]

Data on the relative impact one country's invention technology has on another country is also available in Table 3-3 by examining the numbers in each column exclusive of the main diagonal. The off-diagonal totals for each column are displayed at the bottom of the table. These totals show the relative impact of each country's technology on foreign innovations. In interpreting these totals, only their relative magnitudes are significant. As is readily evident from the table, the United States and Germany have the highest impact on innovations of other countries. This may be interpreted to mean that the export of invention technology from the United States and Germany is strong in both countries; on the other hand, it is relatively weak in France, Japan, and Canada. In contrast, the off-diagonal elements in each row indicates the tendency of a particular country to utilize foreign inventions. As mentioned previously, France is totally self-reliant in invention technology based on the limited French cases analyzed. On the other hand, of the six countries considered, Canada is the most disposed towards utilizing foreign inventions. The data also suggests an interesting insular nature of technology policy in France, where there is high dependence on domestic inventions and little in the way of diffusion into other countries. Germany, however, appears to be quite active both in utilizing foreign inventions and in making their own invention technology available to the rest of the world. Another interesting observation that tends to confirm conventional wisdom is that the flow of invention technology in the United States and Japan appears to be antithetical in that the direction of flow is mainly outward in the former and inward in the latter.

Note that the data in Table 3-3 are derived from a limited sample of successful innovations and are concerned with only the influence of invention technology. Generally, there is a wide range of technological activity that does not result in new inventions; the data should therefore be interpreted accordingly. For example, the data do not show that France actively exports technology in the form of military equipment.[7] Nevertheless, it is instructive to study the flow of invention technology because, to the extent that it reflects each country's technology policies, it provides at least a first-order indication of each country's disposition to utilize foreign technology. On the strength of the available data then, the United States not surprisingly exhibits a net technology outflow; the opposite is true for Japan and Canada. The dependence of Japan on U.S. technology is widely known. Not so well known is Canada's similarly heavy dependence on foreign technology, which has caused the Canadian Government to institute new policies to strengthen its domestic technological capabilities to counterbalance the heavy foreign influence.[6] Of some significance is the evidence that a comparatively high two-way flow of invention technology prevails in Germany. It is the only country of the six that exhibits this balanced picture. Recall that Japan and Germany both exhibit a shorter average innovation time than the United States and the United Kingdom (Table 3-2) The common thread shared by both Japan and Germany from the record of invention-technology data is a strong inflow of foreign technology. Witness, for example, the home video tape recorder originally developed in America but produced and marketed commercially by Japanese firms or the European airbus, which utilizes American (General Electric) jet engines. This leads to the hypothesis that perhaps it is their adeptness in applying foreign developments that gives them the ability to produce successful innovations in a more competitive time frame. Their attentiveness to technological advances in other countries coupled with their disposition to acquire and assimilate appropriate advances to aid domestic innovations may be the keystone for their more timely and efficient process for technological innovation.

A legitimate rejoinder is that despite the shorter average innovation time prevailing in Japan and Germany, the United States still leads in the number of successful and significant innovations. Of the 500 major innovations of the 1953–1973 period, 319 of them originated in the United States. The United Kingdom was responsible for 85, while Japan and Germany had 34 and 33, respectively. Since the large majority of the total was a product of American innovativeness, it could be argued logically that U.S. innovativeness is still unchallenged. However, a closer look at the trend of innovative activity by country over the 1953–1973 period (Figure 3-3) shows an unmistakable slowing of American innovative activity compared to the major Western industrial nations.[8] While nearly 80 percent of the innovations introduced in 1953 were of U.S. origin, by 1973 only about 60 percent of innovations introduced into the market that year came from America. Japan, Germany, and the United Kingdom, on the other hand, managed to sustain moderate increases in their share of major innovations during the

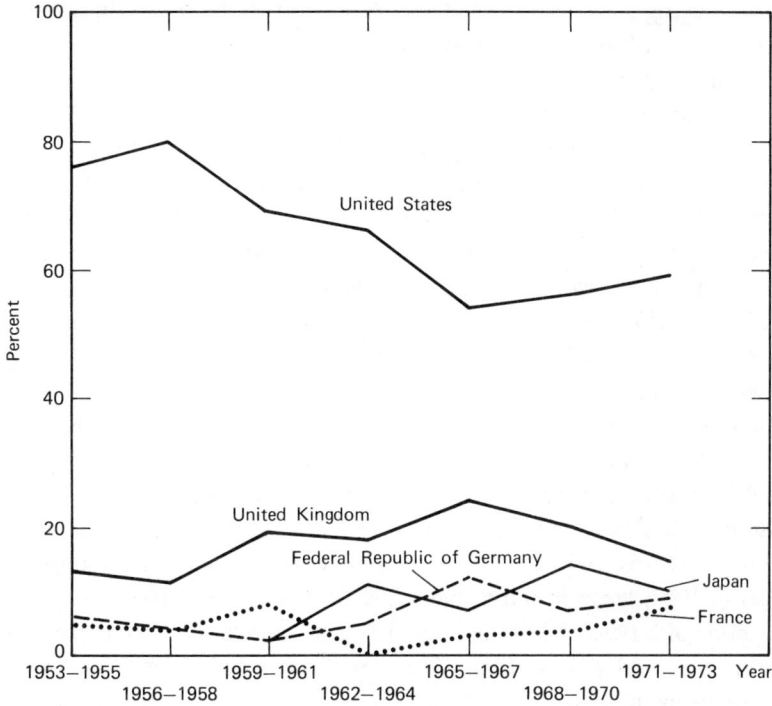

Figure 3-3   Major technological innovations by country. (Source: Reference 3-8)

period shown in Figure 3-3. Japan recorded the sharpest increase, rising from virtually zero during the mid 1950s to a 10 percent figure by 1973.

The erosion of American innovative activity during the period was concentrated mainly in industrial areas where the United States is traditionally strong, namely, the technology-intensive industries, such as scientific instruments, electrical and communications equipment, chemicals and machinery.[8] The technology-intensive areas were also the ones in which the rising share of major innovations from the other countries are concentrated. Japan's innovations were mainly in the machinery industry, and British innovations were generally related to the aircraft industry. Clearly, consistent inroads into markets for technology-intensive products formerly dominated by the United States are being successfully mounted by the major Western industrial nations.

The sagging state of innovation in America may be traced in part to the plight of small business. Because small companies in the United States produce a disproportionately higher share of innovations than do large companies, the recent difficulties encountered by small companies in raising capital undoubtedly contributed in large measure to the decline in American innovativeness. Scarcity of venture capital has discouraged the formation of new technology-oriented firms and hampered expansion plans of established ones. A survey conducted by the National Venture Capital Association showed that there were only 185 new venture financings in 1973 compared to 223 in 1970.[9] The number of small companies able to raise additional equity capital by selling shares to the public has also been declining, as shown by Table 3-4.

In 1969, a total of 698 firms with net worth of under $5 million raised nearly $1.4 billion in public offerings. However, by 1978, the number of small businesses able to do so declined to 21, generating a total of only $129 million in additional capital. In terms of public financing for small, high-technology firms, 204 were able to go public in 1969, but only 4 managed to do so in 1974. Keep in mind that small technology-oriented firms form the backbone of innovation in the United States. Data on the 500 major innovations introduced during 1953-1973 show that small firms produce about four times the number of innovations per R&D dollar as medium-sized firms (1000-10,000 employees) and 24 times as many as large companies.[8]

Fortunately, there are signs that venture capital is becoming more plentiful again. Spurred by the 1978 change in the U.S. tax law where the maximum tax on capital gains was reduced from 49 to 28 percent, venture capital investments are again becoming more attractive to more people.[10] At the same time the tax change has encouraged more entrepreneurs to form new enterprises. As a result, $750 million in investment capital was raised in 1978 for new company formations, an amount that approximates the total raised

Table 3-4   Trend of Public Issues by Small Companies
(Net Worth Under $5 Million)

	Year									
	69	70	71	72	73	74	75	76	77	78
No. issues	698	198	248	409	69	9	4	29	13	21
Share value (millions)	1367	375	551	896	160	16	16	145	43	129

between 1969 and 1977. Additionally, more new issues of small companies are coming onto the stock market, giving venture capitalists greater opportunities to cash out at a profit. As evident in Table 3-4, the number of small companies able to raise equity capital from public offerings has been recovering from the low point registered in 1975.

The cumulative evidence thus presents a picture of eroding American innovativeness relative to the major industrial nations of the West based on a comparison of their respective average innovation period and the trends in national share of major innovations introduced during the years 1953–1973. Greater foreign ability to innovate appears to be related to the foreign countries' attentiveness to technical advances outside their own national borders and their ability to acquire and adapt them to expedite domestic innovations. From this font apparently springs their growing competency in technological innovation, their greater competitiveness in the world markets, and their markedly faster industrial growth relative to the United States as indicated by their higher productivity gains.

Productivity Growth Comparisons

We have seen that technology transfer and technological innovation serves to bridge the gap between the science-technology base and economic progress. Also, the growing innovativeness of the Western industrial nations compared to the United States signifies their growing ability to bridge the technoeconomic interface and to generate greater efficiencies of economic output. In the final analysis, however, it is the effectiveness with which innovativeness appears in the form of higher productivity growth that is important. That this is indeed the case for the leading Western industrial nations can be established by examining their productivity growth rates.

The level of U.S. productivity across all economic sectors still remains the world's highest according to national productivity measures, such as the gross domestic product per employed civilian. Figure 3-4 shows the productivity levels of the major Western industrial nations as a percentage of the U.S. level for the 1960–1977 period.[3] The curves are all indexed to the U.S. level, which is established at 100 in the figure. All countries shown are closing the gap on the U.S. lead by virtue of their faster productivity growth. In particular, Japan, Germany, and France exhibit substantial improvement in their national productivity vis-à-vis the United States during the period. Moreover, there appears to be no immediate change in the trends as the

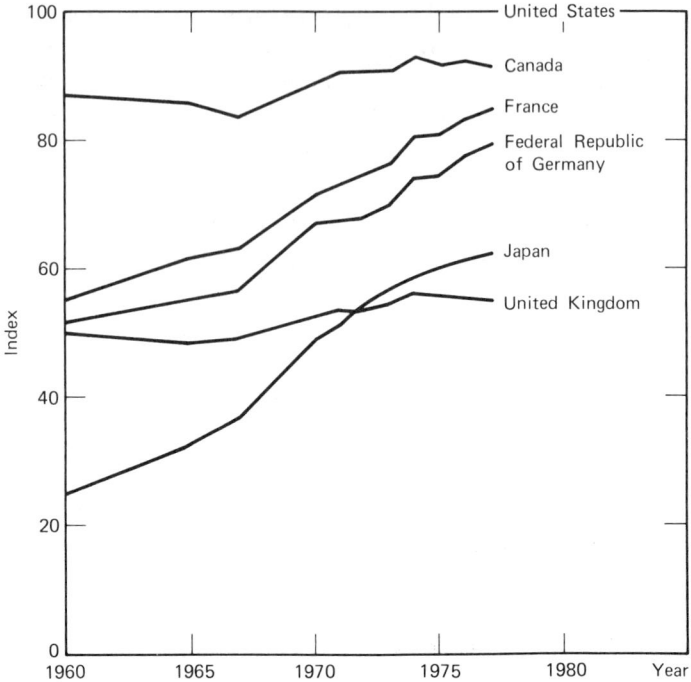

Figure 3-4   GDP per employed civilian for selected countries indexed to the U.S. level. (Source: Reference 3-3)

United States registered a meager 0.4 percent productivity increase in 1978 and a 0.9 percent productivity decline in 1979.

Overall increases in U.S. output per man-hour in the private sector have averaged about 2.5 percent annually between 1870–1965, which exceeded productivity gains in Europe and Japan until 1950. Since 1950, however, the situation has been reversed as higher productivity growth materialized in Europe and Japan as a result of the postwar reconstruction boom.[11] At the same time, productivity gains in the United States has slowed markedly. In the 1947–1966 period, the average annual productivity gain was 3.2 percent.[12,13] However, in the more recent 1967–1977 period, annual productivity increases in the United States slipped to 1.6 percent on average. The slowdown in U.S. productivity since the mid 1960s was fairly widespread, as the majority of industries suffered declines in productivity growth during the period. Notably the U.S. steel industry registered a 1.8 percent annual productivity growth rate for the years 1966–1976. In contrast, the Japanese steel industry's annual productivity gains during the period were several-fold higher, enabling them to surpass the level of U.S. steel productivity by 1976. The productivity picture in the United States was by no means entirely bleak as standout industries such as the electric utilities and the telephone industry managed to register average annual gains of 3.6 percent and 5.6 percent, respectively, during 1966–1976 which, although down from prior years, still compared favorably with the national figure of 1.6 percent. Various projections place U.S. productivity growth to range from 2.0–2.4 percent annually during the 1980s.[12,13] While the pace of growth in Europe and Japan is expected to slow, their growth rates are nevertheless expected to remain measurably higher than in the United States.

Figure 3-5 illustrates the trends in output-per-manhour for the manufacturing sector of select countries during the 1960–1977 period.[3] The curves have all been normalized to their respective 1967 levels for ease of comparison. Only comparative trends are highlighted and not absolute levels on productivity. Clearly, the United States had the slowest productivity growth rate, whereas Japan showed the biggest increases, especially since 1966. The United States recorded nearly a 60 percent gain over the period, while Japan registered a sparkling 279 percent rise. Both Germany and France during the same period registered about a 150 percent gain, while Canada and the United Kingdom boosted productivity by 94.5 and 64.4 percent, respectively, over this period.

A major factor in the slower growth in U.S. productivity is the lower rate of capital investment devoted to upgrading and expanding manufacturing plant and equipment. In the 1960–1973 period, American nonresidential fixed investment averaged 13.6 percent of the national output in goods and services as measured by the GDP. This investment is low in relation to levels of investment prevailing in the major OECD countries for the same period. Japan had the highest investment share of 29 percent. Next was Germany at 20 percent, France at 18.2 percent, Canada at 17.4 percent, and the United Kingdom at 15.2 percent.[14] Comparison of these figures with the average annual productivity gains by country for 1960–1973, depicted in Figure 3-5, shows an unmistakable correlation between investment share and productivity growth. This conclusion is more clearly depicted in Figure 3-6, where the average annual productivity growth in manufacturing industries by country is presented versus capital investment as a percentage of the GDP for the 1960–1973 period. The close correlation suggests that the low rate of U.S. investment in plant and equipment, along with the innovational malaise afflicting the nation, is an important factor in the laggard growth of U.S. productivity since the mid 1960s.

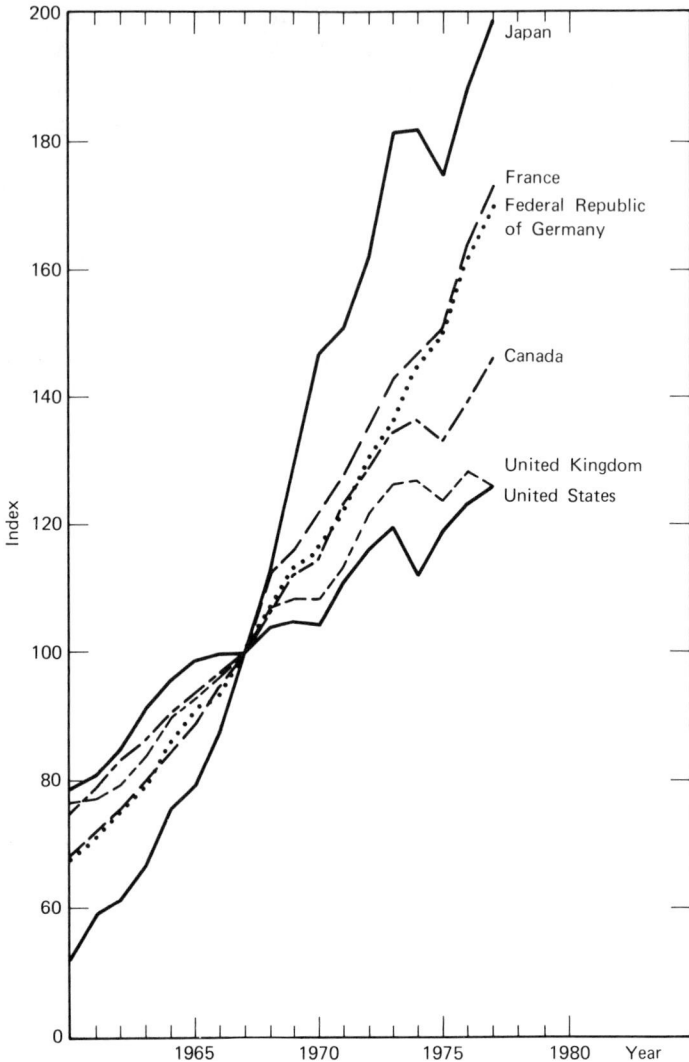

Figure 3-5   Trend of productivity growth in the manufacturing sectors of selected countries. (Source: Reference 3-3)

Moreover, the capital investment trends since 1973 are not very encouraging. In fact, both the level of fixed investment and the proportion of domestic output allocated to investment in the leading industrial nations declined during 1973–1975, precipitated by the global recessionary period of 1973–1974.[15] In addition, direct investments in foreign countries competed

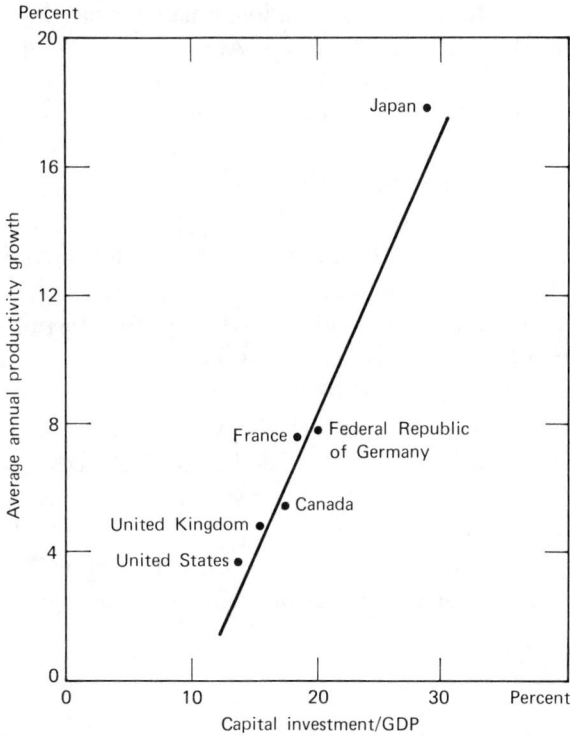

Figure 3-6   Productivity growth versus investment in manufacturing industries by country, 1960–1973.

strongly for U.S. investment capital as the higher profit advantage of investing abroad became more attractive. Meanwhile, modernization and replacement of domestic plant and equipment becomes ever more pressing as the American industrial engine is increasingly stretched past its design limits. Indeed, one estimate places 30–40 percent of U.S. capital equipment to be over 15 years old. Adequate capital formation thus represents a major challenge in maintaining the efficient and competitive manufacturing base necessary to revitalize U.S. productivity growth and maintain competitiveness in international markets.

While an important reason for the lagging growth in U.S. productivity may be the relatively low rate of capital investment in the United States, it is important to realize that the collective wisdom and theories of economists are still insufficient to explain accurately the reasons for the decline in American productivity growth, particularly since 1974. A number of factors are not taken into account by conventional methods of economic analyses, such as the work ethic and attitudes in America, the rise in crime, the increasing national orientation towards a service-based economy, and the effect of more efficient knowledge diffusion.[13] Of these influences many economists rank the ability to transfer and utilize knowledge—technology, management skills, and industrial practices—as even more important than capital investment for boosting productivity growth. Hence although the laggard U.S. productivity growth can not be attributed to any single factor, higher capital investments and a resurgence in technological innovation are widely considered to be major and essential steps for spurring productivity.

The erosion of American inventiveness and technological innovativeness must consequently be reversed to achieve productivity improvements. We need to keep in mind that technological advance alone is insufficient to fuel innovation and productivity growth. Modern and efficient industrial plants and processes plus the ingenuity to transform technological advances into marketable and competitive commercial products are also absolutely essential. Success in boosting productivity growth will eventually lead to more competitive international trade relationships and positive trade balances.

Growing Trade Competitiveness

The most visible evidence of the increased innovativeness and higher productivity growth abroad is found in the strong competition encountered by the United States in both the domestic and international markets. American exports continue to expand but at a slower pace than the growth rate of the world market. As a result the U.S. share of world exports in most categories of international trade has been steadily dropping. The U.S. share of total world exports declined from 14.7 percent to 11.5 percent between 1966–1976.[16] Also, the U.S. share of world exports in manufactured goods slipped by some 20 percent between 1958–1977.[17] The decline in export share of manufactured products, which includes machinery and transport equipment, is particularly significant because this category of exports represents the largest segment of world exports in terms of dollar value. Furthermore, the competitiveness of manufactured exports is indicative of the level of technological know-how and productive efficiency achieved by the exporting nation. Evidently, despite American superiority in many technological fields, U.S. technological advantage is not being effectively translated into competitive goods and services.

One reason American companies do not pursue foreign markets with the intensity of European or Japanese firms is simply that the large size of the American market absorbs most of the attention of American firms. As a result, American firms are not as closely attuned to the intricacies of successful operations in foreign markets. Indeed, many companies may have little interest in foreign markets for a variety of other reasons, such as the considerable expense associated with setting up foreign distribution networks, the red tape involved in obtaining government export licenses, or the necessity of converting manufacturing processes and products to the metric standard.

Other reasons for America's diminished share of the world export market are related to the erosion of technological innovativeness in America. Factors such as the special problems of small, high-technology firms in accessing sources of venture capital and the thickets of government regulations all act to hamper innovation for large and small companies alike. These and other impediments have contributed measurably to lower productivity gains and higher unit costs as more labor input is required just to maintain a given output level. Fuller employment and more costly goods in turn increase the inflationary pressures on the domestic economy, making imported goods all the more attractive. Complex factors such as these all contribute to the inability of American firms to exploit fully their technological advantages, to stem the competitive inroads made by foreign products in the U.S. market, and to compete effectively with foreign firms in international markets.

The increased international competitiveness of foreign enterprises can no longer be attributed entirely to their lower wage scales. Wages and living standards in Western Europe and Japan have risen more rapidly than in the United States since the mid 1960s. By 1975, Western European wages of production workers in the manufacturing sector have generally overtaken those in the United States according to figures released by the U.S. Bureau of Labor Statistics. As shown in Table 3-5, the United States in 1965 had the highest average hourly wage of the nine countries. However, by 1975, the average hourly wage figures in Belgium and Sweden had already risen above the U.S. figure, while in Canada and West Germany the average hourly wage was virtually even with that of the United States. The rising wages abroad reflect not only higher pay standards but also the declining value of the American dollar. As a result, foreign competitiveness is less a consequence of lower labor costs than their growing ability to achieve rapid technology commercialization.

Foreign firms moreover have done a much better job of producing innovations based on American technology. Their success in commercializing technological advances originally pioneered in America has led to a virtually

Table 3–5   Average Hourly Wage of Manufacturing Production Workers in Selected Countries (U.S. Dollars)

	1965	1970	1975
Belgium	1.32	2.08	6.46
Canada	2.29	3.46	6.20
France	1.19	1.74	4.57
Federal Republic of Germany	1.41	2.32	6.19
Italy	1.10	1.75	4.52
Japan	0.48	0.99	3.10
Sweden	1.86	2.93	7.12
United Kingdom	1.13	1.48	3.20
United States	3.15	4.20	6.22

insatiable foreign demand for American technology. A case in point is the field of semiconductor electronics. The United States is the undisputed leader in this field, having produced nearly all the major technical advances from the original discovery of the transistor in 1947 to the latest development in large scale integrated (LSI) circuits.[18] The pace of U.S. technological advance and innovation in semiconductor electronics is unsurpassed. The pocket calculator and the digital watch are examples of successful American innovations in this field. In 1977, sales of digital watches world-wide totaled nearly $2 billion and in a relatively short time has captured one third of the $2.8 billion U.S. watch market. However, for every successful American innovation there are even more successful foreign innovations in consumer electronics, such as miniature radios, television sets, and video tape recorders, that relied on original advances made in the United States. Of course, technology alone is insufficient to assure a successful innovation. Nevertheless, foreign firms seem to be more proficient in the art of transforming technological developments into competitive products.

The West German firm, Siemens AG, typifies this valuable knack for capitalizing on technological advances made by others.[19] Its 1977 sales totaled $10.7 billion with about $1 billion a year earmarked for R&D—which is approximately one eighth of the R&D expenditures for all of West German industry. Instead of allocating the R&D funds for research activities, Siemens applies the bulk of it to product development. Its applications know-how and high technical competence are widely acknowledged, and it employs this strength to good advantage in acquiring and improving on ideas and developments from others. Siemens patented an improved version of Alexander Graham Bell's telephone invention and has developed this into the second largest business in the company and the world's third largest supplier of telephone systems. With American technology acquired in the 1930s plus its own expertise in telegraphy, the company has become a major supplier of telex machines, and in 1977 introduced a new electronic tele-printer that has been selling at the rate of 50,000 units per year. In comput-ers, Siemens employed licensed technology from Radio Corporation of America (RCA) in the 1960s and has since captured a 21 percent share of the mainframe market in Germany, while International Business Machines' (IBM) share has slipped from over 72 to 57 percent. The same strategy is being employed to develop their technological skills in microelectronics and integrated circuits. By means of licensed technology from American firms and exchange agreements with Japanese firms, Siemens hopes to keep abreast of rapid developments in both countries and at the same time de-velop applications in the company's many product lines.

The demand for American technological know-how from companies such as Siemens has never been stronger. In the early years after the last world war, fairly mature technologies were sought by foreign firms to provide them with the manufacturing base to serve their local markets. However, as their technological competence grew and horizons widened, the demand gradually shifted to the more advanced technologies. Increasingly, foreign firms are entering into teaming arrangements with American companies in order to learn of the latest technological advances from their American partners. Generally, both parties in the joint venture benefit from the arrangement; for example, the American partner gains access to new markets in exchange for providing technical and management know-how to the foreign partner. However, the longer-term impact is increased competition not necessarily for the American partner but certainly for the firm's competitors. Consider for example the joint venture between General Electric (GE) and Societe Nationale d'Etude et de Construction de Moteurs d'Aviation (SNECMA), a partially government-owned French aircraft engine manufacturer. The two companies joined to develop and produce a new 10-ton thrust civilian air-craft jet engine primarily for the European market.* The terms of the agree-ment call for GE to supply the hot core for the engine, which is to be drawn from its developmental work on the B-1 bomber. This enables SNECMA to

---

*The fuel economy of the GE-SNECMA CFM-56 engine has also attracted American custom-ers as United Airlines has decided to install them in their DC-8 aircraft.

take a major technological step forward into competition with Pratt & Whitney and Rolls Royce.

Another example is the collaboration between the two leading Japanese computer makers—Fujitsu and Hitachi—and Amdahl Corporation, the American computer firm founded by a former IBM employee.[20] In return for its equity investment in Amdahl, Fujitsu acquired Amdahl technology that made possible the development of their M-series computers in competition with IBM's 370 systems. In January 1978, Fujitsu and Hitachi jointly introduced the M-200 computer, which processes information faster than the fastest IBM model at the time. Not content with this advantage, the two Japanese firms further introduced an improved M-200H model that is even faster than the M-200. IBM has not left the Japanese challenge unanswered, however, as they introduced their new 4300 line early in 1979. Responding in kind, Fujitsu subsequently introduced several additional models featuring high density logic and memory in direct competition with IBM's 4300 mainframes. Thus American technology has helped to herald the entrance of Japanese, firms into head-to-head competition with IBM in a powerful struggle for dominance in the world's computer and electronics markets.

Of course, the joint venture is only one method for transferring advanced technologies, but it is among the more effective ones where the principal parties are near industrial and technological equals. Joint ventures provide opportunities for foreign firms to work alongside and learn from American companies. The advanced technologies acquired along with foreign strengths in product development, applications engineering, quality control, and automation engender keen competition for American industry at home and abroad. The level of competitiveness is particularly high when one considers that the market for technology-intensive products is not necessarily worldwide, but typically is limited only to select industrial nations. Technology-intensive products may be inappropriate for the developing countries because of the inability of the local population to assimilate or utilize it properly. By the same token, less expensive, labor-intensive, mature technologies may be more suitable to the needs of the developing economy than expensive, capital-intensive, labor-saving, high-technology items. Moreover, national security interests may dictate that specific high-technology items may only be exported to U.S. allies. Factors such as these tend to (1) confine the transfer of advanced technologies to the industrial nations, (2) increase the likelihood of developing greater competition in one's own backyard, and (3) raise the level of interdependence between the United States, Japan, and Western Europe.

As the economies of the industrialized countries achieve greater comparability, greater market interdependence develops, Because technology-intensive exports have been the important element in America's trade balance, Japan and Western Europe represent key markets for America's export trade. Robust economies there are necessary to maintain strong demand for U.S. exports. However, Europe has been a shrinking market for American goods for a number of reasons.[21] First, its recovery from the 1973–1974 recession has been slower. Also the high Western European dependence on OPEC oil has eaten into their ability to purchase American goods. Furthermore, the rapid improvement in European technological competence has placed considerably greater competitive pressures on American exports. At the same time, Western Europe is increasingly dependent on exports of industrial goods and foodstuffs to the United States in order to help pay for OPEC oil. These European products then compete with American goods for a share of the U.S. market. As a result, transatlantic interdependence tends to be highly competitive as home markets on both sides of the Atlantic are unable to absorb fully the imports from abroad. The rising tide of competitive imports thus further exacerbates the domestic unemployment problem, prompting calls by threatened home industries for increased protectionist measures from their governments.

Despite the increasingly competitive nature of U.S. trade with Western Europe, the 1978 trade balance was still in favor of the United States by $3.4 billion (Table 3-6).[22] America's trade with Japan, however, has been incurring huge deficits. The deficit in 1978 alone was $11.6 billion, which was an

Table 3-6   1978 U.S. Trade Figures (Billions)

	Exports	Imports	Net
Africa and Asia (excluding OPEC)	18.6	21.9	− 3.4
Australia and Oceania	3.5	2.4	1.1
Canada	28.4	33.5	− 5.2
Europe (Western)	39.9	36.5	3.4
Europe (Eastern)	3.7	1.5	2.2
Japan	12.9	24.5	−11.6
Latin America	17.7	18.7	− 1.0
OPEC	16.6	30.7	−14.1
Totals	141.3	169.7	−28.4

Source: Reference 3-22.

appreciable portion of the total U.S. trade deficit for the year as is evident from the table. This huge trade deficit furthermore understates Japan's huge trade advantage with the United States in manufactured goods because much of America's exports to Japan are in agricultural products and raw materials.[23] The primary cause of this large trade imbalance is not attributed to Japanese protectionism so much as America's lower competitiveness and inadequate attention in promoting exports to Japan.

Japan has instituted a number of measures to reduce the trade imbalance.[24] The government encouraged a stronger yen relative to the dollar, imposed export restrictions on steel and color televisions, increased purchases of American products, reduced tariffs, expanded import quotas and financing, and increased capital spending to develop higher domestic demand. These measures will undoubtedly help to reduce the trade imbalance with the United States. However, the major question is whether a more balanced trade pattern will prove to be only transitory in nature. Working to override efforts for more balanced trade are more deep-rooted differences in the two national economies, such as the significant differential in inflation rates in the United States and Japan. The considerably higher inflation in the United States makes American exports cost more and counteracts what trade advantages accrue from the devaluation in the dollar. Furthermore, the Japanese market economy is heavily export-oriented as opposed to the domestic orientation of the U.S. economy. Regardless of the state of the domestic economy, Japanese firms must continue to produce and export to support their high fixed costs for labor and materials. American manufacturers, on the other hand, are more sensitive to changes in the domestic economy. American exports to Japan also are not rising as rapidly as U.S. manufacturers would like because of the still limited ability of the Japanese economy to absorb American imports. In addition, imports from Europe and the developing countries are competing more and more effectively for a share of the Japanese market. As a result the longer-term prospects for U.S.-Japan trade are still clouded.

A bright spot however is the growing U.S. export trade with developing nations. The share of U.S. exports to the developing nations has been rising rapidly in recent years. During the 1972–1977 period, the proportion of U.S. exports going to the less developed countries rose from 29 percent to 35 percent. Export shares of other Western industrial nations to developing countries have also experienced similar growth. In the same 1972–1977 period, the export share to developing countries rose from 38 to 46 percent in Japan, from 11 to 17 percent in Germany, from 18 to 24 percent in France, and from 15 to 22 percent in Switzerland. Clearly, as the developing nations rush to industrialize, they are becoming export markets of growing importance to the United States and other Western industrial nations.

However, at the same time, the industrialization process in the developing countries will propel them into more direct competition with the industrial nations as they increasingly seek export markets for their newly developed industries. Already, the United States has been experiencing a negative trade balance in nontechnology-intensive goods as items such as steel, footwear, and agricultural foodstuffs enter the United States from Western Europe and increasingly from the developing countries. As the industrialization process in the developing countries continue to unfold, we can anticipate a gradual upgrading of developing country exports into more sophisticated manufactured goods, such as textiles and electronic components, in direct competition with traditional suppliers in the industrial nations. Moreover, some world trade estimates project that future developing nation exports will grow faster than those of the industrial nations. By the year 2000, exports of the developing countries are expected to rise to 27 percent of the world GNP from 17 percent in 1976.[22] Moreover, as new manufacturing plants are built in the developing countries and as the character of their exports increasingly shift from basic commodities to more sophisticated manufactured goods, we can expect to see manufactured exports of the developing nations rise faster during the 1980s than those of the advanced industrial nations. The continual upgrading of developing nation exports signals a most significant shift in future world trade patterns.

The major challenge to the industrial powers comes from the six more advanced developing countries—South Korea, Taiwan, Brazil, Mexico Singapore and Hong Kong.[22] These nations have successfully developed manufacturing industries at a world competitive level. As the process of continual improvement and upgrading of industrial capacities progresses in these countries, the nature of export competition that they present to the United States and other industrial nations can be expected to shift increasingly to more sophisticated items such as steels, autos, and light appliances. In particular, South Korea, Brazil, and Mexico have set their sights on becoming major suppliers of steel before the turn of the century. In addition to basic steelmaking, they also have plans to develop competitive steel-dependent industries such as automaking, shipbuilding, home appliances, and metals fabrication. At the same time other less developed countries can be expected to develop their own manufacturing capabilities in the traditional, less sophisticated industries such as textiles and basic electrical components.

Consequently, as industrialization progresses in the developing countries, the United States along with Western Europe and Japan will find that their traditional export markets will be increasingly eroded as dictated by the comparative advantages enjoyed by the developing countries.

Traditional industries in the United States have consequently sought increased protection from the federal government through the imposition of more restrictive import tariffs and other trade measures. The momentum, however, is in the direction of lower trade barriers, at least among the major industrial nations. The world trade pact completed in Geneva in 1979 after five years of negotiations involving 99 nations will reduce trade tariffs by about one third on $140 billion of annual world trade on some 5700 items over an eight year period beginning in 1980.[22,25] The savings to the American consumer could total $10.6 billion annually as a result of the tariff reductions. More important, the agreement attempts to reduce the proliferation of nontariff trade barriers, such as licensing requirements that could be manipulated to reduce imports, government subsidies that help produce low pricing of product exports, and government procurement standards that favor domestically produced goods. When fully effective the tariff cuts and lower nontariff barriers will mean perhaps $5 billion in added exports annually for the United States.

The Geneva agreement also includes a code that opens certain government procurements among the signatory nations to competitive bidding. This could mean that American companies will be able to bid on as much as $20 billion annually on foreign government procurements that were unavailable to them previously. The chief U.S. negotiator in Geneva estimates that American companies can win as much as $2.3 billion of the foreign procurement market annually, most of which would be in advanced capital goods, electronics, and communications equipment.[25] On the other hand the new code could mean $300 million more in U.S. business for foreign companies. An exception is Japan, which excluded advanced telecommunications equipment from the trade agreement. As a result Japanese firms are still unable to bid on U.S. government procurements. Negotiations between the United States and Japan are continuing, however, to open up about $7 billion of Japanese government procurements to American bids.[26]

Admittedly, the pact acts essentially to lower barriers to trade among the industrialized nations and does virtually nothing to lower barriers to imports from the developing nations. For this reason only one developing country initialed the pact on completion of negotiations. However, the United States has signed separate bilateral agreements with 18 additional developing countries covering both tariff and nontariff understandings. Still, the Geneva trade agreement was notable in that it sets up codes of acceptable behavior on the nontariff barriers. It consequently constitutes an important step towards countering the proliferation of nontariff barriers in world trade.

The answer to the growing trade competitiveness of the developing nations is clearly not to erect even higher import barriers, because doing so would undoubtedly jeopardize current trade relations with the developing countries on which the United States has become increasingly dependent for its exports. Generally, America retains a comparative trade advantage with respect to the developing countries in agricultural products and technology-intensive products. Products in the latter category in fact are required by the developing nations to further their industrialization plans, which eventually will reappear in the form of increased competition for American products. The strategy for the United States then is not to curtail trade but to keep ahead of the developing countries in terms of product specialization, manufacturing know-how, and technological development. As competitive pressures mount on the more mature industries, emphasis need to be gradually shifted to newer, more sophisticated industries in which the United States still retains a comparative advantage. The industrialization process in the developing countries then acts to produce a growing market demand for more specialized goods that American industry must be prepared to meet. The unacceptable alternative is trade wars, which benefit no nation.

The pursuit of complementary export markets by the industrial and developing nations should therefore permit mutually beneficial "North-South" trade patterns to be sustained. The industrial nations in seeking to meet developing-nation market demands for technology-intensive goods and industrial processes can look forward to continued expansion of export markets in these areas. The developing countries in turn can work towards supplying markets in the advanced nations for less sophisticated manufactured items. As the industrial and technological capabilities in the developing countries are progressively upgraded, corresponding upgrading of export markets of the industrial nations can be anticipated. However, industrial nations must recognize that continued imports from the developing countries are important to preserve the purchasing power of the developing world. That is, the industrial nations' drive to expand their own export markets in developing nations can be satisfied only to the extent that developing nation purchasing power is maintained through continues imports of their goods by the industrial nations. The development of complementary export markets for the advanced and developing nations is therefore an important prerequisite for a thriving, mutually beneficial North-South trade relationship.

Another facet of trade competition encountered by American firms is the state-controlled enterprise (SCE) found in both industrialized and developing countries alike.[27] SCEs are endowed with distinct advantages over the privately owned American firms in international competition. National governments provide their SCEs with guaranteed markets, tax advantages, and interest-free loans. In return the SCEs may continue to operate inefficient plants to sell products abroad at a loss in order to maintain employment, gain market share, and to earn foreign exchange. Politically, it is advantageous for SCEs to be export oriented because it keeps the labor rolls full and helps the nation's trade balance. Examples of SCEs are British Steel in the United Kingdom, Italsider (metal refining, steel) in Italy, Renault (autos) in France, and Salzgitten (steel, shipbuilding) in West Germany. However, it is in the developing countries where rapid growth in the number of SCEs has come. The SCEs generally are involved in raw materials and manufacturing and not so much in consumer goods basically because of their lower strategic importance as viewed by the central authorities. As a result, basic industries requiring large capital investments, such as steelmaking, are widely populated by SCEs.

The SCEs have posed formidable competition to American firms as attested to by the current state of the steel, textile, and shoe industries in the United States. Protectionist measures have been adopted for these industries in the form of government subsidies and import restrictions. However, while the basic industries bear the brunt of the SCE competition, technology-intensive industries are not immune to it. SCEs from the industrialized East and West alike, such as Britain's ICL, (International Computers Ltd.) in computers or Rolls-Royce in aircraft engines, are pushing technology-intensive exports in direct competition with American firms. In fact, competition for American firms in the civil aviation field comes entirely from SCEs. Even Brazil with their state-controlled aircraft company, Embraer, is competing with American manufacturers like Beech Aircraft. The competitive picture seems to be brightening somewhat for American industry, however, as some nations appear to be reassessing their drive towards a nationalized industry because of the considerable costs involved in subsidizing inefficient plants and low-priced exports. A number of American firms meanwhile have found it advantageous to enter into joint ventures with SCEs, such as the GE/SNECMA arrangement discussed earlier and the Boeing/Rolls-Royce deal. In the latter, Boeing uses Rolls-Royces engines in its 757 aircraft. Subsequently, Boeing obtained a $600 million contract from British Airways for 19 of the 757 aircraft.

There is little doubt that SCEs will remain very much on the international trade scene. Although politically expedient, government protectionism in response to the SCE challenge is not the best answer as it serves only to raise instead of lower the barriers to trade. Arrangements such as the Boeing/ Rolls-Royce deal may become increasingly necessary as American companies seek to accommodate themselves to the competitive advantages enjoyed by the SCEs in the world marketplace.

Changing NATO Trade Relationships

America's international trade relationships are undergoing fundamental change not only in the commercial area but also in relation to trade in military equipment and armaments among the nations of the North Atlantic Treaty Organization (NATO).* Considerably more than just commercial interests are at stake here as the major issues concern the collective security of 15 sovereign nations of Western Europe and North America. Still, the economic health of NATO member nations is closely linked with alliance security interests and military capabilities to meet the threat posed by the Warsaw Pact nations.‡ The United States traditionally has been the principal supplier of conventional arms and equipment for NATO forces. However, a vigorous defense industrial base has gradually arisen in Western Europe, stimulated in the main by American technological and economic assistance during the postwar period. As a result America's traditional role as equipments and armaments supplier for NATO is being strongly challenged by the growth of technology-based defense industries primarily in the United Kingdom, France, and West Germany. Keen competition for weapon system sales to the NATO military market has contributed to the proliferation of nonstandardized equipment. The transatlantic competition engendered has subsequently produced protectionist tendencies not unlike those encountered in the commercial markets. At issue is the European drive to increase their share of the NATO defense market. Apparently, only an

---

*Formed in 1949, the members are Belgium, Canada, Denmark, France, Federal Republic of Germany, Greece, Iceland, Italy, Luxembourg, The Netherlands, Norway, Portugal, Turkey,the United Kingdom, and the United States.

‡Established in 1955, the member nations are Albania, Bulgaria, Czechoslovakia, German Democratic Republic, Hungary, Poland, Rumania, and the Soviet Union.

express U.S. willingness to accommodate a greater European participation in the NATO defense market would forestall the Europeans from closing their markets to the United States even if American equipment proves superior or cheaper than their own. The Europeans seek a "two-way street" in transatlantic defense trade, where the objective is to achieve a more balanced flow of military equipment and armaments between North America and Western Europe. The undesirable alternative is a foreclosure of America's major export position in the North Atlantic defense market.

The European industries' drive to expand their defense markets is understandable in view of the heavy financial investments they have already made in building up an important defense industrial capability. Of course, national pride is also a factor as are the economic and political benefits accruing from an expanding job market. To continue to thrive, the European defense industries must seek to widen their market base. Nowhere is there a larger defense market than in the United States. In defense R&D alone, the annual expenditures in the United States total some $12 billion, which is nearly three times the amount spent by all NATO allies combined,[1,28] but the American defense market has generally been a protected market reserved for domestic companies. Although legislation such as the Buy-American Act has been effectively waived in the interests of promoting transatlantic cooperation, many additional difficulties remain to confront European companies. The sheer size of the North American defense market compared to that presented by the individual nation-states of Europe poses problems of scale and diversity for European firms. Development and production of weapons systems on a combined European-American scale is generally beyond the present-day capacity of European industries. Trying to serve the American defense market also requires incisive knowledge of the U.S. government procurement system coupled with the fortitude and nimbleness to pursue government contracts in open competition with American firms. The process is unlike that in Europe where the smaller national defense markets are not able to support many competing companies and where procurement awards typically result from advance government-industry collaboration. Additionally, U.S. government policies concerning foreign access to classified information as well as critical technologies must be reexamined and revised in order to accommodate European bidders on military procurements that require access to classified data.

A legitimate concern on the part of American industry is whether reciprocal measures will be undertaken by European governments to open up government defense markets for fair and open competition by American firms.

The typical European practice of making contract awards without undergoing competitive bidding plus their various buy-national policies have acted to discourage potential American bidders on European defense contracts. Whether a true quid pro quo actually develops in opening the transatlantic defense markets to open competition remains to be seen. It is important to note, however, that the 1979 Geneva world trade agreement includes a code on government procurement. It provides that government procurements among all the signatory nations will be opened to competitive bidding. However, the code does not apply to government procurements that involve national security.[29]

How does increased competition from European firms in the North American defense market contribute to a stronger NATO? This is a logical question when we observe that the prime motivation for the "two-way street" in transatlantic defense trade stems from the European desire to expand markets for national defense industries in the direction of the North American continent. The question is particularly important when one considers that the rise in national competition for the NATO military market prompted the present proliferation of nonstandardized equipment and weapons in the NATO arsenal. The lack of standardization within NATO severely limits combat effectiveness and flexibility of force deployment in a real conflict. For example, there are 24 different types of aircraft in the NATO air forces.[30] This diversity of aircraft inventory hampers NATO's overall ability to refuel, rearm, and service these aircraft. It has been estimated that the use of standardized equipment would improve combat effectiveness of NATO forces by 30–50 percent and up to 300 percent in some air units.[31] These estimates are indicative of the appreciable improvements in combat effectiveness possible from deploying standardized or, if nonstandardized, at least interoperable equipment and armaments.

Despite the acknowledged industrial strength of the major NATO nations, the lack of standardization in NATO material is symptomatic of an inefficient system for military systems development and procurement within the Atlantic Alliance. North America and the European Community represent the two most technologically advanced industrial economies in the world; yet the Warsaw Pact countries are distinctly superior in terms of producing and deploying standardized conventional arms and equipment. Rather than engaging in duplicative development and production of competitive systems, the Warsaw Pact nations are designing and producing to common requirements. As a result they are able to field military forces employing large quantities of standardized arms and equipment.

The issue then is not so much whether or how to eliminate competition in development and production of military systems among the NATO countries, but rather how to best harness the benefits of competition within an all-encompassing framework for armaments cooperation. An open, properly managed, competitive procurement system will usually produce the best there is to offer. On the other hand, uncoordinated and excessive competition is wasteful and divisive. It leads to inefficient use of national resources arising from duplicative development of competing systems and fragmentation of markets unable to support full production-scale economies. Excessive competition furthermore exacerbates internecine conflicts that jeopardize the ability of the NATO Alliance to fulfill its mission for collective security effectively. However, the formula by which national aspirations may be subordinated to the common defense has proven to be highly elusive.

The basic problem lies in the dissimilarities of the national economic systems on both sides of the Atlantic. Rather than a single integrated NATO market there exists 15 separate national markets each with its own set of procurement regulations and differing market accessibility. To deal with the problem the United States initiated a three-pronged approach. The triad of cooperative actions involve negotiating bilateral memoranda of understanding (MOUs) in reciprocal purchasing, dual production, and the Family of Weapons concept for weapons development. The general MOUs are intended to improve mutual cross-national market accessibility by removing governmental "buy-national" restrictions on a reciprocal basis and thereby foster fair competition by NATO defense contractors in the defense markets of each member nation. In this way each NATO country can seek out the most qualified and competitive source for their respective defense procurements. General MOUs have already been negotiated with the governments of Canada, West Germany, Italy, Norway, the Netherlands, and the United Kingdom. The second leg of the triad is dual production, which means setting up two or more weapon production sites in different NATO nations. The aim is to reduce, if not eliminate, duplicative R&D prior to production. Under this concept, a single nation will perform the necessary R&D. After completion of development, two or more production activities will be instituted in different countries to avoid the trade and labor imbalances that would result from exclusive production and sales. A significant experience base has already been acquired on cooperative production programs such as the ROLAND II missile system, the BULLPUP missile, and the F-104 aircraft.[32] The third leg of the triad is the Family of Weapons concept. The idea is to achieve higher efficiency in weapons development by dividing up the responsibility for developing weapons having similar missions among different nations. In this manner greater coordination and cooperation in R&D expenditures might be achieved within the Alliance and needless duplication of effort minimized. Transatlantic industrial teaming is encouraged whereby a nation having lead responsibility in a particular weapon can save a portion of the development work to be performed by overseas subcontractors. Each nation will fund the program for which it has lead responsibility.

The degree of success achievable from these U.S. initiatives will depend heavily on how forthcoming the European allies will be in taking reciprocal actions. Unless reciprocal steps are taken on the European side, the goal of a common NATO defense market will not be achieved. It must also be recognized that progress towards greater cooperation at the government level is still insufficient to assure a common NATO defense market. Important disparities also remain at the industry level. In the United States industry is accustomed to open competition in a customer-oriented market economy. Competition dictates product quality and price levels. The large size of the American market is able to support a number of competitive enterprises; at the same time, it reduces reliance on export markets. On the other hand, the character of European markets differs markedly. They tend to be socialistic economies heavily influenced by government laws affecting the welfare of workers.[29] The benefits of competition become diffused in an economy where much of industry is nationalized. Government procurements are awarded to European contractors by means of "chosen-instrument" selection instead of open competition. This characteristic is an outgrowth of the smaller scale of European national markets, which can not support large numbers of competing firms. Small market size has also generated greater dependence on export markets to maintain production levels and to preserve jobs. In order to penetrate the much larger American defense market and to compete effectively with American firms, the Europeans have joined in many multinational programs to assemble the industrial strength heretofore found only in the United States or the Soviet Union.[33] Political and administrative procedures have been established to pool R&D and financial resources allowing Europeans in essence to scale up to the American defense market. European consortia have in fact produced the Airbus, ROLAND missile, Alpha jet, and the multirole combat aircraft (MRCA) among others.

The differences in the industrial make-up and markets on both sides of the Atlantic constitute a significant mismatch in scale and structure. Transatlantic teaming arrangements between American and European enterprises must somehow reconcile and accommodate these differences. An important consideration in multinational programs is the greater administrative burden and longer program lifetimes. European experience with multinational co-development programs indicates that generally the developmental time associated with multinational programs could be up to twice as long for complete development of a particular weapon system as the developmental time required by comparable national programs. Potential savings in R&D costs associated with multinational codevelopment projects could in fact be appreciably offset by higher administrative costs as well as the higher costs stemming from longer developmental times. One would therefore reasonably expect that higher administrative and developmental costs would also occur in transatlantic industrial teaming arrangements for codevelopment and coproduction. The degree to which these added costs would offset R&D savings would have to be determined project by project, but it would probably not be insignificant. However, operational advantages associated with the eventual use of common weapon systems and equipment derived from codevelopment and coproduction will generally be the deciding factors in any attempt at cost-benefit analysis.

How to accommodate both competition and cooperation on an international scale is not a simple task. Although the combat forces from NATO's member nations have been successfully integrated under NATO military command, the same is not true in the materiel procurement field. There is no integrated NATO competitive procurement system.[34] Herein lies the crux of the problem. The lack of a NATO-wide procurement system with established policies and guidelines for competitive selection of armaments and equipment exposes Alliance procurement practices to the nonuniformities of widely different national procurement systems and policies. International cooperation in competitive procurement must be founded on a common base of mutually agreed guidelines and procedures by which all parties can abide. Otherwise, conflicting national interests and selection procedures will inevitably dominate and divide. The first step towards a solution could be to assign some of the responsibilities of a weapon system procurement agency to a NATO organization. Its mission would be to develop Alliance procedures for competitive procurement of combat equipment and weapon systems. It would also provide guidelines for the formation of transatlantic industrial consortia seeking to respond to NATO defense procurements. Eventually, national differences in procurement policies and regulations could possibly be reconciled and protective barriers removed or waived in the interests of NATO cooperation. Government defense markets on both sides of the Atlantic might then be opened to fair competition under a common Alliance procurement system.

The two-way-street concept has been promoted as a means by which NATO standardization can be achieved. The argument is that by opening up the government defense markets on each side of the Atlantic and building a balanced defense trade in both directions, closer Alliance cooperation will be developed; this will ameliorate the more divisive influence of competition and lead to more nations adopting common equipments. Unfortunately, the two-way-street concept has been but a nebulous concept subject to differing interpretations with no real effort to define the relevant terms of reference. A major question is whether the concept embraces only the armaments trade or includes the total trade in armaments and defense-related goods and services. The European view is that the traffic count on the two-way street should be confined only to the armaments trade inasmuch as it is the category in which the trade balance heavily favors the United States. However, if the total defense trade between the United States and Europe is considered, including indirect, nonarmament items such as U.S. expenditures for supporting ground forces stationed in Europe, the trade balance becomes heavily weighted in Europe's favor.[34] To establish a balanced transatlantic armaments trade, European NATO nations look for reciprocal U.S. purchases of their goods to "offset" their purchases of U.S. armaments. However, offset purchases by the United States have taken the form of defense-related nonarmament items which, other than to satisfy the quid pro quo precept of the two-way street, may be totally unnecessary and not cost-effective and may not contribute at all to furthering NATO standardization. This also points to the inconsistent interpretations given by Europeans as to whether nonarmaments count on the two-way street. Expediency appears to be the major determinant. As a result the U.S. Department of Defense has established the policy of not entering into offset agreements unless it is absolutely necessary to complete a transaction considered to be of significant importance to national security.

There appears to be no real reason why only a segment of the total defense trade should be considered as part of the two-way street other than that it provides a convenient rationalization for greater European penetration into the U.S. defense market. If this is the case, then the prime motivation for the Europeans is one of economics and market access rather than improved Alliance security through armaments cooperation as in the case of the United States. The popular European rejoinder is that being able to sell more to the United States strengthens European defense industries and by so doing contributes to a stronger European partner in the Alliance. Hence a divergence in priorities and perceptions seems to exist in America and Western Europe concerning the rationale for a balanced transatlantic defense trade.

Consider for example the F-16 aircraft coproduction project involving the United States and European coproducers Belgium, Denmark, the Netherlands, and Norway. A range of motives characterized the participation of the four European partners in the project.[35] The prime consideration appeared to be economic, as the prospect of more jobs was an important factor, especially in Belgium and the Netherlands. Potential technological spinoffs was also a major factor for the participants. In particular, the Norwegians cited the prospect of technological gains that could possibly be diffused throughout their defense and civilian industries to be of primary importance. It is interesting to note that achieving NATO armaments standardization was not mentioned as the major consideration by the European coproducers. The reason is that pursuit of NATO standardization was not a significant factor during the pre-project negotiations. The prime consideration even for the United States at the time was market access and foreign sales.

The two-way street is also not without its American detractors. Military arguments against it point to an undesirably heavy American dependence on European sources for arms, equipment, and spare parts that would be subject to interruption in time of war or political estrangement. Risks associated with increased American dependence on foreign weapon systems also concern their reliability of performance, compatibility with American-made weapons platforms, timeliness of supplies, and cost escalation. Efforts to reduce these risks will tend to offset potential cost-savings derived from nonduplicative development and production. Economic arguments against U.S. purchase of European weapon systems cite the potential loss of American jobs and corporate profits. Both labor and the U.S. defense industry consider these sacrifices too expensive a proposition for furthering a concept of allied cooperation weighted in favor of the Europeans.

Even if, as may very well develop, a more balanced two-way trade materializes in spite of the apparent divergence of American and European motivations, the trade flow itself may resemble more a multilane highway rather than a two-way street. The reason is, of course, that trade negotiations must still be conducted and concluded at the national level. Without a collective NATO or European procurement system, transatlantic defense trade will continue to reflect primarily bilateral interests founded on government-to-government agreements without necessarily representing the best interests of the Atlantic Alliance as a whole. Under these circumstances national competitive forces will remain dominant over international arms cooperation in the collective interest. Even a multilateral arrangement, as in the case of the F-16 project, has come under criticism from other European NATO nations mainly because the project helps to set up competing aircraft production plants in the participating European countries despite the fact that existing aircraft production facilities lie underutilized in other European countries.

International armaments cooperation and NATO standardization could probably be better achieved through collective European efforts rather than from separate U.S. bilateral agreements with individual European governments. A major restructuring and consolidation of European institutions and policies is necessary to aggregate European interests in the Atlantic Alliance. Close cooperation in defense trade presently exists between the United States and Canada. In order to forge a similar, mutually beneficial, transatlantic defense trade between North America and Western Europe, a process of "Europeanization" is needed to consolidate the diverse national interests and resources of the European member nations. Positive steps have been taken in this direction through the formation and activities of the Independent European Programme Group (IEPG). The IEPG is composed of the European member nations of NATO and was organized basically to develop a stronger and more cohesive European defense contribution to the Atlantic Alliance.[36] Formed in 1976, the IEPG appears to be making good progress towards strengthening European cooperation within the Alliance. Further harbingers of increased European cooperation are the different European consortia formed for specific development projects. For example, a British, West German, and Italian consortium was formed to produce the Tornado MRCA. Another example is the ROLAND II missile, which is the product of a French-German joint development. The momentum then is unmistakably in the direction of closer European cooperation.

From the U.S. perspective, recent moves to establish a two-way alliance trade beneficial to defense industries on both sides of the Atlantic is not a license for the uncontrolled export of American technology. Precautionary measures must still be observed to ensure that technology releases are consistent with sound business practices and established U.S. regulations and policies. In the case of the F-16 project, several categories of critical technologies were identified. They were removed from coproduction consideration and supplied to the European partners in the form of "black boxes." However, eventual release of the critical technologies may yet occur on a time-phased basis when security restrictions permit.

The F-16 example helps to point out a number of difficulties and contradictions involving technology transfers with NATO allies. First, developing a two-way transatlantic alliance trade may contribute to NATO weapons standardization, but it also means giving up a portion of the American defense market to European competition. Second, transfers of key technologies to NATO nations strengthens the military capability of the Alliance but at the same time may serve to weaken the U.S. defense industrial competitiveness in the world market relative to European defense industries.[37] Third, American companies are continually faced with the contradiction of selling just enough technology to secure a business deal without making the foreign companies into viable commercial competitors. Fourth, transfers of technology overseas may also jeopardize U.S. technological leads in the military sphere so that transfers must be tempered in accordance with U.S. export control policies and guidelines. Fifth, the distinct time sensitivity of technology dictates that technology-release options must undergo continual review and evaluation as individual technologies retain singular obsolescent rates. A time-phased release of technologies is often the most practical approach.

As a final note, it is important not to let current preoccupations with technology export controls overshadow measurable benefits that may come to the United States in acquiring and utilizing technology from abroad. In light of their rising technological competence and rapid industrial progress, foreign countries should be viewed as an additional resource to complement the American science-technology base. Furthermore, foreign experiences and institutions for ready commercialization of domestic and imported technologies may hold important lessons for the United States that should not be ignored.

CHAPTER **4**

Foreign Technology as a Resource

In order to be more competitive in the international market, America must recapture the innovative vitality that characterized its performance immediately after World War II. The rapid emergence of Western European and Japanese competition is more a consequence of foreign resourcefulness and dedication to rebuilding after the war's devastation than America's largesse in giving away technology. Of course, superior American technology contributed in no small way to postwar industrial reconstruction in Western Europe and Japan. However, the essential ingredient was foreign ingenuity and their dedication not only to rebuild but also to improve on the prewar situation. Consider, for example, the foreign steel industries. In rebuilding after the war, foreign steelmakers were quicker to install new and more efficient processes, such as the basic oxygen furnace, on a wider scale than American steelmakers. It is this more innovative disposition that has quickly led Europe and Japan to challenge American markets and, more recently, American technological leadership. Hence the principal challenge for America is recapturing its own former innovative instincts.

More protectionism and—to the extent that national security is not jeopardized—tighter controls on U.S. technology exports are obviously not the answer. The main argument for slowing the outflow of American technology abroad aside from national security considerations is that foreign firms are taking advantage of American-financed R&D to produce new innovations that compete with American products. The International Association of Machinists and Aerospace Workers in 1977 claimed that Japan has reaped $15 billion from the purchase of $1.5 billion worth of American technology.[1] The fact is that this same American technology was also available to American companies. The difference lies in the ability of foreign enterprises to make better use of American technology. Thus the answer lies in developing ways to use available technological resources more effectively to produce more innovative and competitive goods rather than in raising higher the protectionist curtain. Thus American technological horizons should not be limited only to domestic technology but should also be expanded to include foreign-developed technology. Indeed, it could well be that foreign technology could be as helpful to the United States as U.S. technology has been to Western Europe and Japan.

The case for devoting more attention to foreign technological developments is compelling. While the proportion of national resources allocated towards the generation of new technology by means of research and development has been declining, the demands for technology from the economic, social, and military-political sectors have been inexorably rising. On the economic front, inflation, high unemployment, declining productivity growth, and deficits in the trade balance have been particularly conspicuous in the 1970s. Technological progress is expected to lead the American economy back towards recovery. Associated with the economic demands for technology are strident social demands associated with the need to conserve energy, preserve the environment, provide for urban mass transportation, produce adequate housing for the poor, and so on. On the military-political front increased reliance on technology is required to help protect the nation's security. Consider for example the second U.S.-USSR Strategic Arms Limitation Treaty (SALT II). A vital issue for America, particularly after the loss of the Iranian listening posts, is treaty verification, or the ability to monitor and to detect treaty violations by the Soviet Union. Whether the latest, most sophisticated electronics gear at the remaining listening posts will be able to detect unauthorized testing of strategic weapons over the entire Soviet land mass is crucial to U.S. security interests.

The conflicting circumstance of higher economic, social, and military-political demands for technology at a time of declining national investment in R&D is further compounded by the competitive outgrowth of the rising foreign technological challenge. The situation has prompted renewed efforts to develop more effective use of the domestic science-technology base. The technology-oriented federal agencies, most notably the Department of Defense and the National Aeronautics and Space Administration, have instituted domestic programs to develop in both public and private sectors effective mechanisms and institutional arrangements for realizing greater benefits from our nation's investment in science and technology. However, undue preoccupation with the domestic science-technology base represents at best a suboptimum strategy in the face of the incipient shifting of the world technological balance overseas.

Comparable attention needs to be focused on foreign technological developments and their potential for import and adaptation in order to help fuel technological innovation at home in response to the rising expectations of the socioeconomic and military-political sectors. Technology in foreign industrialized countries is rapidly growing to be a potentially valuable resource to complement domestic technology and therefore should not be neglected by the United States. Rather than viewing the improved technological capabilities of foreign industrialized nations as a threat or challenge to the United States, foreign technology should be viewed as a potentially valuable asset. What better way to meet the foreign technological challenge in the face of scarce domestic R&D resources than by importing, adapting, and utilizing their technology to help meet priority needs at home? In this way adversity may be changed to advantage in an otherwise unpromising scenario.

Of course, foreign technology will not supplant domestic R&D. Instead, domestic R&D should be complementary to foreign technology in the sense that it may be used to improve or to adapt foreign technology to domestic needs. The commitment of R&D resources in the United States should reflect adequate prior consideration of available foreign alternatives. Thus rather than displacing the need for R&D, cognizance of foreign technological developments for potential acquisition presents an additional option for R&D planning and decisionmaking. It could provide the added flexibility needed for more efficient use of limited R&D resources and more effective response to problems of high national priority.

Foreign experiences in monitoring technological developments in other countries, importing appropriate technologies, and applying them to help solve socioeconomic and military problems may also hold valuable lessons for the United States. Considering that the United States in recent years has been operating more in the role of a supplier than an importer of technology, the United States can reasonably be viewed as basically a novice compared to countries in Western Europe and Japan when it comes to exploiting foreign-developed technology. Developing insights into the institutional processes employed in different Western industrial countries for tracking American technological progress and adapting them to local needs would be exceptionally valuable in helping the United States align its own governmental and industrial processes to better utilize foreign technological resources. Additionally, increased awareness of the criteria and important factors involved in decisions to acquire technology from abroad as opposed to proceeding with domestic R&D programs is fundamental to a more flexible posture for technology utilization. It is granted that methods used and experiences gained by foreign nations in this regard will probably not be directly transferable to or emulated in the United States because of institutional and cultural disparities inherent in different societies. Still, lessons gleaned from the experience of others cannot fail to be useful in helping to determine governmental policy options available to the United States in promoting the adaptation and use of foreign technology.

Certainly, the potential usefulness of foreign-developed technology to the United States is amply demonstrated by numerous innovations already familiar to the American public but which, though less well-known, nevertheless depended on foreign technology. The jet engine for aircraft, the more efficient basic oxygen process for steelmaking, auto disc brakes, steel-belted radial tires, and polypropylene plastic widely used in American cars and appliances are innovations commonly found in the United States that originated from abroad.[2,3] These are but a few of many more foreign innovations that contribute to the American economy and quality of life. Yet there is scarcely any organized American effort to keep up with foreign technological developments on which innovations such as these are based.

The U.S. Department of Commerce in a draft study on U.S. technology policy states,[4]

> No Government agency is responsible for the continuing assessment of foreign technology developments in noncommunist countries. This omission contributes to present export controls inadequately protecting national security and economic interests that involve critical design and manufacturing technology.

The military services do have offices in London and Tokyo to monitor scientific activity in Western Europe and Japan, respectively, but their efforts are narrowly focused with inadequate attention being devoted to technology applications and industrial progress. There are scientific attachés at selected U.S. embassies overseas but their duties are to keep abreast of science and technology policies in the host countries that might impact on U.S. foreign policy and international relations. They are generally not concerned with gathering information on technological developments and industrial progress for potential exploitation in the United States.

Foreign countries, however, have set up listening posts in a number of American cities expressly to gather information on American science, technology, and industrial development. Their embassies generally are staffed with technically trained personnel keeping tabs on American industry. The British have in fact gone a step further in setting up a state company in the United States expressly to purchase American technology and to entice technical specialists to relocate to Great Britain.[5] Parallel efforts are also being made by representatives of foreign companies in the United States. Foreign efforts are greatly assisted by the relatively open access to U.S. technical information and technology afforded them in the United States. Unclassified U.S. government scientific and technical publications are offered for sale through the National Technical Information Service (NTIS) of the Department of Commerce. Sales to foreign customers have done well; they amounted to nearly 17 percent of total sales in 1978.[6] NTIS therefore provides foreign customers with a convenient one-stop shopping center for America's scientific and technological progress. Obtaining samples of American technology is almost as easy; for example, foreign representatives could go to the nearest electronics outlet and purchase a variety of microprocessors for shipment home. Learning about microprocessors is made even easier with the large numbers of public seminars being offered on how to program and apply microprocessors. The only entrance requirement is the cost of registration.

A marked difference therefore exists in the inclination and ability of the United States vis-à-vis other countries to acquire imported technology. The United States has been slow to recognize the potential benefits of a positive and active effort to acquire and to utilize foreign technology. Other countries in contrast find the relatively open access to American science and technology a veritable gold mine that they are panning aggressively. Tangible evidence of the overall effectiveness of foreign efforts is found in the high imbalance of transnational license fees in favor of U.S. receipts for technology licensed by foreign enterprises. Yet it behooves the United States to recognize that the incipient shifting of the world technological balance overseas dictates also a concomitant shift in U.S. technology policy towards greater sensitivity to and exploitation of foreign technology.

The Promise of Foreign Technology

Foreign countries remember the miracles of technologically based industrial innovation in the United States dating back to the Industrial Revolution and have witnessed the major role that American technology played in the economic resurgence of West Germany and Japan since World War II. These historical lessons have not been lost on them but have since found expression in their determined efforts to adapt imported technology to local needs. The developing countries in particular realize that, in order to reap the full benefits of an increasingly technological world, they must align their governmental, commercial, and educational infrastructure for effective utilization of technology imported mainly from the industrialized countries. The lessons of history are viewed from an entirely different perspective in the United States. The natural sentiment is a feeling of pride in American technological supremacy; this feeling has been the wellspring for much of America's modern self-reliance and confidence in its own technological capabilities. It has not, however, evoked a wide appreciation for the potential benefits that could develop from a more condescending attitude towards foreign technology. Yet the potential advantages from importing and adapting foreign technology to meet domestic needs can be substantial, especially in light of the rising technological competence abroad. As is already well appreciated in other countries, imported technology can provide an important option that should not be neglected by R&D planners and managers in the United States.

One may argue that importing foreign technology will certainly not help the nation's trade balance. On the contrary, the acquisition of foreign technology, for example, through licensing agreements or the purchase of manufacturing rights, would increase the net U.S. dollar outflow because of the payment of royalties and fees to foreign concerns. However, this adverse effect would be appreciably offset by the savings in R&D investment that would otherwise be needed to develop a similar technology. Another important factor is the possibility that refinements or improvements in the imported technology might provide the needed competitive edge for the recipient company to expand from the domestic into the international market. The resultant increase in export volume could more than compensate for the adverse effect on the U.S. trade balance caused by the acquisition and use of the foreign technology. A not insignificant political consideration is the added contributions imported technology from Western Europe could have on furthering transatlantic technological cooperation within the NATO alliance. To the extent that the imported technology is in the form of advanced weapon systems, it would also serve to produce a more balanced two-way armaments trade between the United States and Western Europe.

An important consideration in the acquisition of foreign technology is that the process is highly problem- or need-specific. Typically, a well-defined problem or need exists for which a technological solution is sought. The options are either to seek and apply technological solutions developed by others or to initiate research and development to try to solve the problem locally. The acquisition of technology developed by others being highly demand responsive means that decisions to do so usually are carefully weighed and reached only after extensive and comprehensive search and evaluation of candidate technologies for technical and economic appropriateness. As a result, the likelihood that the technology finally selected will meet the problem requirement is much higher than when relying on one's own R&D inasmuch as the results of R&D are subject to many uncertainties, and the probability of eventually reaching successful application is low. The risk-reward ratio associated with the acquisition of foreign technology therefore is considerably lower than that generally available from domestic R&D.

The savings in R&D investment from choosing the imported-technology option could furthermore turn out to be a substantial proportion of the acquisition costs for foreign technology. In fact, R&D costs could very well exceed acquisition costs where the necessary R&D entails high technical risks or expensive capital equipment. Also, as is widely appreciated, the successful outcome of any R&D project cannot be assured. The acquisition of external technology, on the other hand, usually involves proven technology with low technical risk. The decision by the industrial manager to launch a new-product development could thus be based on firmer technical judgments if already developed technology is used instead of sole reliance on internal R&D. The industrial manager can also do a better job planning for the effort since acquisition costs can be more accurately identified than R&D costs, which are notoriously difficult to predict. The use of foreign technology would also provide the R&D planner with added flexibility in the allocation of scarce R&D resources, allowing him to concentrate resources on R&D projects that would complement foreign technology to best advantage. This flexibility is especially important in an era of declining R&D investment. Finally, the quicker availability of proven technology acquired from abroad could produce significant time savings, which are all-important to the industrial manager in a highly competitive environment. Hence, the promise of (1) lower technical risk, (2) more flexible resource allocation, (3) more accurate planning, and (4) potential time and cost savings should constitute powerful incentives for the industrial manager to increase utilization of foreign technology.

The importance of these considerations is, of course, not limited to the industrial scene. They are very much areas of foremost concern to government planners and program managers. In fact, the potential benefits from foreign technology may be more widely recognized within the government by reason of recent Department of Defense directives that require that alternative foreign systems and solutions be adquately considered in the defense systems acquisition process. Thus major weapon-system development programs of the military services must first show that no alternative foreign system exists before the development program can proceed. Even after a developmental program is underway, the acquisition of foreign technology could contribute significantly to problem solving in the different phases of the program—from R&D to engineering development and even into production. The net effect could be to permit the program to achieve planned milestones faster or to enhance measurably the operational capabilities of the system under development.

In the NATO context, greater sensitivity to foreign technological developments would make it easier for codevelopment and coproduction projects to be established between the United States and NATO allies. The effort to develop more transatlantic cooperation in armaments development and production is the bedrock on which current policies to promote weapons standardization and interoperability rest. Armaments cooperation with allied nations becomes most imperative in view of the fact that in some areas, such as aircraft and tactical missiles, the leading Western European countries are considered technologically equal to the United States. Codevelopment and coproduction is intended to arrest the proliferation of competing systems within the Atlantic Alliance and to provide a framework within which cooperative R&D and production might be established in order to make most efficient use of allied resources. Substantial cost savings are anticipated from less duplication in development and production programs within the Alliance. However, the greater administrative burden encountered in multinational codevelopment and coproduction programs is expected to partially offset anticipated cost savings as well as lengthen the time required for program completion. In the F-16 aircraft coproduction project for example, the main problems encountered were those of management and administration rather than of technology. Still, if codevelopment and coproduction projects with European allies are to be encouraged, more precise information on allied technological and production capabilities is required.

Substantial social benefits are also possible from a more receptive stance towards foreign technology. The new power plant in Saugus, Massachusetts for generating electricity from garbage would not have been possible without licensed technology from Europe. Many of the pollution control devices available in the United States are also derived from European technology. In the area of technology and planning in public transportation, both Japan and West Germany are considerably ahead of the United States. In fact, the United States would probably benefit substantially from large infusions of West German and Japanese technology in public transportation.

In Japan, the best known example is the Japanese National Railways' Shinkansen bullet trains, which cover the 310 mile Tokyo-Osaka run in little more than three hours.[7] On an average workday these trains carry more than 350,000 passengers on 235 train trips. The load factor is even higher on weekends. Moreover, the bullet trains have been very profitable. The entire construction cost of the Shinkansen has been paid fully in the first eight years of operation. Currently, bullet-train service is being expanded to cover an eventual 4200 mile network. However, it is not just the Shinkansen that is notable in Japan. The Japanese suburban railway and city rapid transit systems are also excellent. Tokyo is widely recognized to have the most advanced rapid transit system among the cities of the world. Safety records of Japanese trains are outstanding, and rarely do train collisions involving passenger injury occur. The high popularity of the Japanese transportation system stems from the safe, dependable, and convenient service it provides to the public. Clearly, the United States can learn much from the Japanese experience and innovations in public transportation systems. Much of their technology could undoubtedly also be adapted to improving the present U.S. transportation system and facilities. Moreover, if the U.S. transportation system approaches the Japanese excellence in service, more Americans will be induced to leave their automobiles and use public transit systems. The social benefits could truly be far-reaching in terms of reduced traffic congestion in the cities, cleaner air from less automobile-exhaust emissions, and slower growth of gasoline consumption in the nation.

Perhaps the most important outgrowth of a greater disposition by Americans to use foreign technology would be in helping to stimulate more innovative thinking through more familiarity and closer interaction with the foreign technological community. Personal relationships with foreign peers gradually built up over time would promote the exchange of ideas and viewpoints, which is important in the idea-generation process. New ideas tend to be more forthcoming in a mutually interactive environment. Because new ideas are the source of new innovations, stimulated idea generation at home from increased awareness of technological developments abroad would ultimately enhance America's ability to innovate. Equally important is the discovery that existing ideas and approaches have already been tried by others and were found wanting. In these circumstances the discovery could precipitate a redirection or reorientation of a particular innovation into more fruit-

ful avenues. It could even dictate the termination of the innovation. In either
case, unnecessarily large investments of money, manpower, and materials
will have been avoided. The resultant effect would be to improve the efficiency of the technological innovation process in American by increasing the
likelihood of success. Combined with a stimulated ability to generate new
ideas, a stronger rate of technological innovation in the United States could
therefore be anticipated.

The Japanese Experience with Imported Technology

One need not look any further than Japan for a modern day testament to the
importance of imported technology in spurring technological innovation
and economic growth. In fact, Japanese society is itself a deep blend of
things foreign and domestic. As far back as the seventh and eighth centuries,
the Japanese acquired technology and the foundations of their national language from the Chinese.[8] After several years of isolation from the outside
world, the Japanese in the fifteenth and sixteenth centuries resumed importing technology from China and several European countries. Firearms were
introduced during this period. After another period of germination and internal development, the transfer of scientific knowledge and technology
from abroad was revived in the nineteenth century as an important instrument for achieving the Japanese goal of catching up with the West both
economically and militarily. It was this third wave of imported technology
during the Meiji Restoration that brought Japan from a feudal state to an
industrial and military power in the years leading up to World War II. As
military and political events came to a head, technology transfers from
abroad necessarily gave way to domestic technology development immediately prior to and during World War II.

  After the war Japan's national objective was a total commitment to reconstruction of its war-torn economy. With only a few exceptions, Japan found
that it had to import technology from abroad again in order to acquire the
needed know-how in the shortest possible time. This dependence on foreign
technology was brought about by the particular circumstances prevailing in
Japan after the war. Much of Japanese technology developed prior to and
during the war was subsequently found to infringe on patents owned by
foreign enterprises. As a result, Japanese companies necessarily had to depend on foreign enterprises for licensed technology. Furthermore, as the
demilitarization process unfolded in postwar Japan, new companies sprang

up to serve the domestic market. Japanese engineers fresh from military development projects quickly adapted themselves to working on commercial development projects. Competition was strong, and time was therefore of the essence. Japanese companies had neither the time nor the financial resources to develop their own technologies without infringing on existing foreign patents. Their only recourse was to license or purchase technology from abroad. The acquisition of technology from abroad consequently became the core of the nation's technological efforts as they worked to rebuild their national economy and to catch up with the Western industrial nations. This emphasis on utilizing foreign technology is mirrored in the rapid rise in Japanese companies' licensing activity with foreign enterprises. In their 1973 fiscal year (ending in March) more than 1931 licenses for foreign technology were concluded, compared to only 142 in the 1952 fiscal year.[9,10]

Japan's active efforts to acquire foreign technology after the war did not come at the expense of domestic R&D. On the contrary, domestic R&D was, and is, selectively employed to complement technology from abroad. It is directed towards improving or adapting the imported technology to fit specific applications. A market and applications orientation characterizes much of the R&D effort. Japanese R&D is directed towards meeting specific goals and industrial or commercial applications. A look at the record of Japan's investment in R&D confirms the importance attached to a strong domestic R&D program to complement imported technologies. Figure 3-2 showed the comparatively strong upward trend in Japan's R&D investment as a share of their GNP during the 1962–1974 period. In contrast, the U.S. R&D investment was in a distinct downward trend over the same period. Considering that Japan's GNP has been the fastest rising among the Western industrial nations during the postwar period, it is clear then that its investments in R&D has also been rising in absolute terms at an unprecedented rate. In the 1960 fiscal year, Japan's R&D investments amounted to only $500 million, but by fiscal year 1972, their R&D expenditures reached $5.3 billion. Additionally, the composition of their R&D expenditures is heavily oriented towards the industrial sector with very little devoted to military R&D. Typically 70 percent of the total R&D is industrially funded in Japan, whereas in the United States only about 45 percent of total R&D comes from the industrial sector. Of the remaining 30 percent government-funded R&D in Japan, about three fourths of it relates to the civilian economy, whereas the United States devotes only about 40 percent of government R&D to civilian-related projects. Thus the overwhelming majority of Japanese R&D expenditures are for projects intended for direct civilian and industrial use.

Japanese R&D investment, moreover, is expected to continue climbing as the nation becomes less dependent on imported technology to fuel continued technological progress and to maintain the pace of industrial growth.

Japan's committment to a strong domestic R&D program to complement foreign technology has figured prominently in their industrial development and economic growth in two important ways. First, it has enabled them to shift priorities and concentrate limited resources in areas of increasingly high value-added production to sustain the rising per capita income of its citizens.[11] This continuing restructuring of their economy is necessitated because of shortages of labor and raw materials and rapid changes in markets and industrially relevant technologies. In the years immediately after World War II Japan's industrial development centered on labor-intensive industries such as cotton textiles, agriculture, and machinery in order to recover from the war's devastation and to provide employment for its labor force. Gradually, the emphasis was shifted in the 1950s to more capital-intensive industries producing more sophisticated products such as automobiles, petrochemicals, machine tools, and electronics. This shift to more sophisticated products coincided, of course, with their mounting ability to innovate based on imported technology. Subsequent Japanese commercial successes in fields such as consumer electronics, automobiles, shipbuilding, railways, general and precision machinery, and electrical equipment are well known.[9] Another redirection of Japan's industrial policies appears to have been introduced in the 1970s as priorities seem to have shifted towards the high-technology industries such as computers, telecommunications, aircraft, aerospace, and nuclear energy. As more resources are shifted into industrial areas of greater economic promise and technological sophistication, it must be recognized that it is Japan's resilient R&D infrastructure, built on a rich blend of domestic R&D and foreign technology, that provides it with the means to penetrate new, high-growth markets.

Domestic R&D has also contributed considerably to Japan's strong position in international markets. It has been the wellspring for product and process innovations as well as for improvements in imported technology from which a competitive export trade has developed. Japan's export trade is essential to pay for its high dependence on imported raw materials, agricultural products as well as technology. Being a resource-poor country, Japan is 85 percent dependent on imported energy resources. It must import raw materials and basically recycle them through its technologically oriented industries into higher-value manufactured goods for the export market.[11] Contrary to popular opinion, however, Japan's domestic market represents the primary market for its industries. Products exported abroad typically represent "surplus" production over and above what the domestic market can bear. Production for export is maintained in order to keep the plants operating near capacity, to keep unit costs low, and to produce higher profits from domestic sales. Their pricing policies on export goods are intended to capture and hold market share in order to maintain an outlet for surplus production. Consequently, domestic R&D is the key ingredient in the Japanese cauldron of indigenous skills mixed with imported technology and raw materials that enable them to produce highly competitive goods for the export market.

Examples of Japanese proficiency in applying their technological skills to produce internationally competitive innovations are legion. To name a few, consider their tape recorders, hi-fi equipment, motorcycles, watches, cameras, pianos, and snowmobiles. Their original application of the transistor for pocket-sized radios is also well known. While the United States was pioneering in transistor research, Japanese engineers were busily learning and applying their knowledge in building the first commercial transistor and transistorized radio. From this modest start, a multibillion dollar consumer electronics industry in Japan eventually developed. In the early years, however, Japanese products acquired a not-undeserved reputation for being shoddily made and of low quality. To their credit, however, they quickly grasped the fundamentals of value engineering and quality control, again learning mostly from the Americans and partly from the Europeans, and they proceeded to absorb and apply them in industrial practice.[12] The results of these efforts are well known as Japanese products have since achieved a reputation of high quality and good responsiveness to demands of the marketplace.

Japan has been a major customer of the United States in semiconductor electronics for many years. In what has become typical fashion, these imports are recycled in value-added products with the addition of large doses of domestic R&D, ultimately reappearing in the form of advanced integrated circuits and powerful mainframe computers that rival the best that IBM can produce. Their goal is to surpass the United States in semiconductor technology and achieve world dominance in the computer market in the 1980s.[13] To this end, they have initiated a $250 million government-industry VLSI program to develop the necessary advanced technologies and manufacturing techniques. The intensity of their efforts to acquire American semiconductor technology has not diminished but rather has escalated markedly. Not content with conventional licensing and product imports, Japanese firms are entering into an assortment of arrangements with American firms, such as equity participation as in the Fujitsu-Amdahl deal, sponsoring semiconductor R&D projects in American firms such as Fujitsu's long-term R&D contract with Microtechnology Corp. of Sunnyvale, California, and the establishment of new Japanese-owned companies in America intended to tap American semiconductor know-how, as in the case of Japan's Matsushita setting up its American research laboratory, Microelectronics Technology Corp., in Palo Alto, California. All these diverse avenues are employed to speed the transfer of U.S. semiconductor technology to Japan in order to achieve global supremacy in computers and semiconductors within their planned time frame.

As the Japanese develop further their technological skills in semiconductors and work with their characteristic singleness of purpose to challenge IBM, they have had to rely increasingly on the export market to lower production costs of integrated circuits and computer subsystems. Competitive costs on semiconductor components and computer subsystems are essential if mainframe computers themselves are to be competitive. Because of the high R&D and capital costs associated with semiconductor R&D and computer production, planning and marketing strategies must necessarily be directed at the world market. Japan can no longer expect to export their surplus production in semiconductors and computers but must work directly to increase their share of the world market. As a result their overseas shipments of semiconductor products have progressively grown in both scale and sophistication since their entrance into the export market in 1973.[14] Most of their exports go to other Asian countries, but an increasing share is being sold to the United States and Western Europe. Manufacturing operations have also been established both in the United States and Europe to get around anticipated protectionist measures as well as to have greater access to technological know-how in the competitor countries.

One such company is Omron R&D, Inc. located in Mountain View, California.[15] The company was set up in 1970 as a wholly owned subsidiary of Omron Tateisi Electronics Co. of Kyoto, Japan for the purpose of tapping U.S. technological know-how and developing new electronic products for the American market. Its product lines include electronic calculators and computer equipments. American scientists and engineers make up the technical staff, but the company is run in the typical style of Japanese companies, which differs considerably from management philosophies found in American companies. Instead of heavy emphasis on profits as in American companies, product diversification and future growth is stressed. Also, an indulgent system of personnel management typical of Japanese companies is employed. The meshing of Eastern management style and corporate goals with Western culture and values have not been smooth but becomes no less important as more Japanese firms establish beachheads in America and Europe.

Certainly there are other areas besides computers in which Japan in making rapid strides. One such area is in robotics. Japan is the acknowledged leader among the Western nations in the manufacture and use of industrial robots for automated production. A total of 30,000 units has already been installed in Japan compared to only about 4500 in the United States and all of Western Europe. The Japanese understand well that the importance of robotics to the economy lies not in being just another new, high-growth area but also in potential contributions to raising national productivity from increased automation of industrial processes. Another area is solar energy R&D. Because Japan must import nearly all its energy fuel stocks, effective harnessing of renewable energy resources is an urgent priority. Solar water heaters are already in mass production and common use. Solar energy conversion of seawater to produce hydrogen fuel is under investigation. Their expertise in mariculture also provides important insight into solar energy conversion for producing chemical feedstocks as well as organic matter or methane gas to power industrial plants.

Japan's extraordinary record of economic growth during the postwar period testifies to their success in bridging the technoeconomic interface. Several factors appear to be key to their success. First, the political climate in Japan after the war has been one of high stability. With the dismantling of the military establishment, the tightly knit government, industry, and financial sectors were able to concentrate on economic development as the top national objective. With unswerving singleness of purpose but yet recognizing their resource limitations, Japan pursued an industrial strategy that concentrated available resources in high-growth industries where technological progress bore the promise of higher productivity growth. Japan's technology policy is completely identified and fully integrated with achieving the nation's economic goals. The emphasis is on the commercialization of technology rather than on its creation. Research and development in Japan is concentrated in the industrial sector to expedite innovations and to achieve technological advantages over competing enterprises. Unlike America, which is preoccupied with domestic R&D, the Japanese display the knack for making productive use of modern technology wherever it may be found. The Japanese outlook that the world is one vast schoolroom is certainly one that Americans would do well to adopt. Not the least important factor is the existence of an industrious and well-educated labor force that displays, by Western standards, intense company loyalty and a predilection to get along with management. The notable self-discipline of Japanese workers has been a major force in the high labor productivity in Japan despite the higher wage demands of recent years which, with the devaluation of the dollar, by 1978 had boosted average Japanese wages higher than those in the United States.[16]

Within the space of 100 years the close coupling of technological progress to economic growth has transformed Japan into the world's foremost economic power today according to economic yardsticks such as productivity growth, industrial exports, and GNP growth. Political power has grown along with economic power as Japan's voice in international councils is given greater weight and as its influence in foreign affairs spreads. However, detracting from Japan's extraordinary economic growth are persistent domestic problems confronting Japanese society. To Japanese consumers unacceptably slow progress has been made in pollution control, traffic congestion, safety, health, consumer protection, and welfare. Belated recognition has been given to the need for technological solutions in these social areas as well. Progress is being made. In pollution control Japan's emission standards for new industrial plants and automobiles are the most stringent in the world. As a result, the most serious pollution sources have been measurably reduced since the early 1970s when the pollution control problem first gained prominent attention. That these nonindustrial problems have been addressed only recently points to the heavy emphasis placed on economic development in the past and the comparatively lower priorities assigned to social needs. If societal problems are attacked with the same vigor as economic problems have been in the past, future improvement in the quality of life for Japanese citizens would seem to be assured.

The Japanese experience in technology-based economic growth is generally difficult to apply to the American scene because of the many differences in culture, traditions, and the political and social forces shaping each country's national goals and policies. Typical of the distinct differences in the two countries is the divergence in corporate goals in the private sector. Where American companies are in business to make profits, Japanese companies consider profits secondary to growth, market share, and contribution to social progress. Nevertheless, Japan's enviable record of higher productivity growth has caused Americans to take a closer look at, and in some cases even adopt, Japanese management styles and programs. More major U.S. corporations are studying, experimenting, and emulating the Japanese methods for participative management in trying to boost worker productivity through stronger worker identification with the corporate organization. The major American auto companies—General Motors, Ford, and Chrysler—have all initiated some changes in work patterns to more resemble the Japanese production methods. The U.S. Department of Commerce is also planning a "cooperative technology program" whereby government and industry would join in cooperative projects designed to expedite industrial innovation and productivity. This program is patterned after Japanese cooperative laboratories like the one set up between the Japanese Ministry of International Trade and Industry (MITI) along with five of Japan's computer manufacturers to try to crack American dominance in large-scale integrated circuits.[17]

The lesson then is not that the United States should emulate Japan's policies and practices but that Japan's economic resurgence has been built on a cornerstone of foreign technology, a source that has been relatively neglected in America to date. The Japanese example is illustrative of the giant economic strides possible when an enlightened technology policy, closely attuned to economic goals, seeks to exploit fully the technological resources available both at home and abroad.

Why the Indolent U.S. Approach to Foreign Technology?

The natural question is why Americans are slow to capitalize on foreign technological developments. Probably a major factor is the fact that the erosion of U.S. technological leadership has been manifest only in recent years, and we have simply not had time to adjust our policies and accustomed methods of operation to accommodate the changing world technological balance. The achievements of the nation's scientific and engineering establishment have been widely recognized in modern times. Examples of America's technological supremacy abound, such as the space spectacular of placing the first man on the moon. Achievements of this kind naturally contribute to an introspective perspective and pride in the nation's technological prowess. The virtues of self-reliance on technological matters have furthermore been ingrained in the typical scientist and engineer from the early days of his college training. Also, the markets for America's technological talents have been confined largely to the domestic front. However, the fact remains that over 60 percent of scientific and technical publications in the world originate outside the United States.[18] To the extent that this is a representative measure of world-wide scientific and technological output, it is reasonable to expect that more than half of progress in science and technology comes from outside the United States. In spite of foreign spectaculars such as the launching of Sputnik I by the Soviet Union in 1957, it takes a long time before recognition and acceptance of foreign technological competence begins to take hold within the American engineering community. It takes an even longer period of time for this acceptance to be manifest in more aggressive policies and procedures to utilize the results of foreign technological activity.

American investments in R&D have typically been among the highest in the industrial world. However, the budget figures are somewhat misleading when one considers that roughly 60 percent of government R&D is spent on defense and space.[18] Although the nondefense and nonspace R&D budgets have grown rapidly in recent years, most of the growth has been in the health, energy, and environmental fields with only a small proportion of these budgets being earmarked for R&D with direct economic relevance. Other industrial nations of the Free World, however, allocate a greater proportion of their R&D resources to projects having industrial value and a smaller share to defense and space. Therefore, a fundamental mismatch in national R&D priorities seems to exist between the United States and its allies; this further adds to American perceptions that foreign technology is not readily adaptable to U.S. needs. This perception is reinforced by the popular notion that progress in U.S. military and space programs are not readily transferable to the socioeconomic sphere. The closed-loop procurement system employed by both the Department of Defense and the National Aeronautics and Space Administration (NASA) features detailed technical requirements, performance specifications, and a guaranteed market for the end product. No such clear-cut definition of technical requirements or delineation of market structure exists in the commercial sector. Notable civilian spinoffs of military developments such as numerically controlled machines and the Boeing 707 commercial aircraft are considered more as exceptions than the rule. However, both the Defense Department and NASA have instituted concerted efforts to transfer their respective military and space technologies to the civilian sector. Limited successes achieved so far have helped to moderate the notion of limited transferability of military and space technologies.

Moreover, the implications of a mismatch in American R&D priorities with respect to those in Western Europe and Japan may be overstated. The fact is that Japan and other European countries have not found the order of U.S. R&D priorities to be significant impediments to their efforts in applying American technology and know-how for their needs. Technology in itself may find many applications in contrast to the technological end product, which only serves the purpose for which it was designed. Identification of the key technology stripped of its product embodiment then constitutes the important step in minimizing the influence of disparities in national R&D priorities. The mismatch in national R&D priorities itself concerns only the national R&D budgets of the United States vis-à-vis those of other counties. Nevertheless, there still exists considerable commonality of interests and objectives among the commercial sectors of the different countries. Yet American scepticism in pursuing foreign-developed technologies extends across both public and private sectors. Thus there must exist more fundamental reasons for America's slowness to exploit foreign technology more fully.

The lack of clear-cut government policies in this area to provide incentives and assistance to American industry has not helped. By and large, industry has been left to fend for itself in respect to keeping abreast of foreign developments. The stock rationale given is that normal free-market mechanisms are sufficient to motivate industry. Where this laissez-faire attitude of government may have been suitable when American markets were secure and technologies essentially unchallenged from abroad, it is becoming less so in today's more interdependent world of greater international trade competitiveness and rising foreign technological competence. Especially in the case of American firms competing with foreign SCEs could government assistance help to reduce significant competitive advantages enjoyed by the foreign SCEs.

Fortunately, more forceful government leadership is becoming evident in encouraging greater utilization of foreign technological resources. This is especially true in relation to our NATO allies. In order to encourage greater standardization and interoperability of weapons and equipment, the military services are required to consider alternative foreign systems and resources before development and procurement of new weapon systems can proceed. Where foreign systems are adequate, they are to be acquired in lieu of domestic development. In addition, the 1979 Geneva world trade agreement opens up more government procurements on both sides of the Atlantic to foreign bidders. American R&D procurements have always been available to foreign bidders, but U.S. bidders have preference unless the foreign source can provide a distinct price advantage or unless it possesses superior technology. What is significant today, however, are conscious efforts by the government to encourage more foreign sources in defense procurements both as prime contractors and subcontractors. In particular, U.S. policy on the handling of classified data is being reformulated to accommodate more participation by foreign sources in defense procurements. These efforts therefore underscore the clear determination on the part of the Carter Administration to cooperate genuinely with European allies and to utilize foreign assets more effectively.

However, American industry is not exactly enthusiastic about the indicated policy trends. The move to utilize more effectively foreign technological assets is widely viewed as giving up a substantial portion of the U.S. government procurement market to competing foreign enterprises. The attendant loss in business, jobs, and profits cannot be clearly rationalized in their minds. The government argues that only in opening up U.S. government procurements to foreign bidders will European allies also open up their respective government markets to American firms. However, this argument is suspect from the American industry perspective because it basically displaces a large, assured U.S. government market for an uncertain, more fragmented, and therefore more risky, European defense market. Furthermore, there are strong doubts that American firms will be subject to open and fair competition with European firms many of whom enjoy close, chosen-instrument relationships with their respective central governments. As a consequence, American industry has not rushed to embrace President Carter's NATO initiatives or his concept of arms cooperation with the European allies.

There are even more fundamental reasons why American industry has not been aggressively pursuing technology from abroad. The large American commercial market has produced a strong domestic business orientation within a large segment of American industry. As a result, international opportunities have suffered from benign neglect. Granted, many of the large, high-technology firms do maintain staff to track foreign developments, but most of this effort is confined to the company's competitive area and the information gathered is not widely disseminated. A few enterprising exceptions like GE have initiated subscription services where data gathered on foreign technological activities are available to subscribers on a continuing basis. The fact remains, however, that the large majority of American firms do not regularly monitor or appraise developments in other countries. Thus the domestic character of much of American industry has led to low export consciousness and a decided inattentiveness to foreign developments.

Furthermore, the typical large industrial organization in the United States is not structured to foster technological innovations, let alone innovations based on foreign technology. The reason lies in the much longer time required to bring an innovation to market as compared to the typically shorter time horizon of the industrial manager. An innovation can take anywhere from 3–30 years, and sometimes even longer, for successful realization. As was previously pointed out (Table 3-2), the average innovation period in the United States during 1953–1973 was 7.4 years. Of course, many more innovations never reach successful realization, but are terminated for a variety of reasons. The odds against an innovation being successfully commercialized are perhaps 50 to 1. Therefore, the risks are high. However, the typical large company is operated to minimize risk and to increase profits. The performance of the industrial manager is gauged by the annual return on investment achieved by his operating unit. He can ill afford to invest in risky new ventures that, even if successful, will not start generating returns for many years. The fundamental incompatibility between the industrial manager's preoccupation for short-term gains compared to the high risk and longer time horizons in the innovation process serves to discourage innovation in more mature companies. It is therefore not surprising that they devote limited attention to foreign technological resources.

In fact, there is evidence that the large industrial organization pays less attention to external—both domestic and foreign—sources of technology than do the smaller firms.[19] Figure 4-1 illustrates the proportion of U.S. innovations (among the 500 significant innovations of the 1953–1973 period) that employed external sources of technology as a function of firm size. Companies with more than 10,000 employees tended to depend less on external sources of technology than did the smaller companies. Aside from the low orientation towards technological innovation in most big companies, their lower reliance on outside technology sources undoubtedly stems also from the comparatively larger internal resources available to them. Medium-sized firms of 1001–5000 employees showed the lowest dependence on outside technology sources; small companies of 1–100 people showed the high-

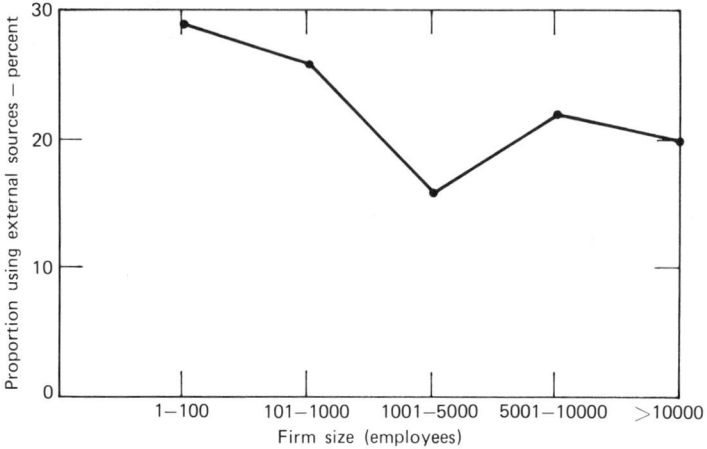

Figure 4-1   Utilization of external sources according to size of firm.

est dependence. It is also significant to note from the figure that over 70 percent of U.S. innovations relied on in-house technology only. Considering also that outside technology will generally come from domestic rather than foreign sources, it is clear that very little use of foreign technology developed as part of America's innovational efforts.

In contrast to the big companies, small companies tend to be more innovative in terms of producing more major innovations per R&D dollar, and they rely more on external sources of technology.[18,20] However, a completely different set of barriers serves to thwart their active use of foreign technology. Small companies must necessarily be innovative to survive against their larger competitors. They are therefore less hesitant about utilizing outside technology if it is necessary to provide a better, more competitive product. Their greater outward orientation also stems from their much smaller technological and financial resources base. Licensing technology from outside is often cheaper than maintaining a sizeable R&D staff and investing in risky R&D projects. However, small companies, must restrict their technology acquisition to items that do not require overly heavy investments of capital and manpower as to unduly strain their limited resources. Small companies furthermore generally do not have the technical staff necessary to keep abreast of outside technological developments let alone to try to track foreign developments. This particular activity is a luxury generally reserved for big companies that small companies can ill afford. Small companies also tend to be even more domestic oriented than big companies. Because of the large size of the American commercial market, small manufacturers generally try to serve local or regional markets. Whereas acquiring a share of the national market may pique the ambitions of many of them, serving the foreign market generally requires much larger corporate assets than typically available to the small company. As a result, small manufacturers tend to concentrate on the domestic market and are not particularly sensitive to foreign developments. Since 97 percent of America's factories are small manufacturers employing less than 500 people,[21] a logical conclusion is that the American industrial base as a whole is not well attuned to technological developments abroad.

Individual attitudes also determine to a large extent American propensity to acquire and use foreign technology. An important consideration is the "not-invented-here" (NIH) factor, which leads people to resist using outside technology. In many organizations a professional value system exists that assigns higher status and rewards for originality and inventiveness than for the application of technology developed by others. This attitude is deeply ingrained and difficult to reconcile with technology transfer and utilization of foreign technology when it has been encouraged throughout the educational and professional life of the individual. A value system that assigns more weight to innovational achievements would be a source of encouragement for individuals to seek technology and solutions wherever possible rather than immediately developing their own "optimized" approach. After all, it is the ability to relate seemingly disconnected and diverse factors and to weave them into an elegant response to a problem or need that lies at the heart of success in technology transfer and innovation.

A less condescending attitude and greater recognition of the market potential and possible long-term benefits derivable from foreign technology should be adopted by American innovators. Rather than viewing a particular item of foreign technology in terms of its deficiencies and limitations, the perspective should be one of how it could be modified or improved to provide a better product or how it could be adapted towards an entirely new market demand. In deciding whether or not to enter into a joint venture with a foreign partner, consideration should be given not only to the short-term commercial benefits realizable from the new venture but also to the longer-term technical benefits that might be gained from such a partnership. Also, purchasing or licensing foreign technology should not be considered an admission of a company's own inadequacies in the particular technology area but should reflect credit on the organization for having (1) the technical sophistication and awareness to recognize the technological and commercial implications of the foreign technology and (2) the fortitude and foresight to acquire the technology in order to expedite a particular innovation. As more opportunities arise to deal with foreign enterprises it becomes more important to seek truly two-way technological exchanges instead of the familiar one-way flow out of the United States typical of the past. After all, losing technological leadership by one country means there is more to be learned from other countries.

Prospects for a Revitalized Outlook

The storm clouds of slower economic progress at home and keener trade competition abroad may yet have a silver lining. In the same manner that the launching of Sputnik I awakened America to pursue and eventually overtake the Soviet Union in space exploration, so too would we hope that America's economic difficulties of the recent past would help spur a revived effort to meet the technological and economic challenge from abroad. Indeed, the foreign technological challenge is becoming more evident to more Americans. That this increased awareness might eventually lead to an effective response has yet to be demonstrated. Still, there are increasing signs that technological and economic developments in foreign countries are being elevated in the American consciousness.

The U.S. government through the Department of Commerce has for years sought to promote American export trade through promotional efforts such as trade exhibits abroad and widespread publicity at home. So far, however, the efforts have not turned the tide of deficit trade balances. The appropriateness of U.S. technology policy for correcting the trade imbalance and laggard industrial productivity growth has been investigated and found wanting. The lack of an organized government effort to track and assess technology in the Free World has also been decried. The need to promote exports and develop better information on foreign technology has more recently found expression in the domestic policy review of industrial innovation initiated by President Carter in 1978. Drawing on the views of business, industry, labor, academicians, and the American public, the Advisory Subcommittee on Patent and Information Policy among its recommendations voiced the need for the U.S. government (1) to assist the private sector, especially small businesses, in learning of trade opportunities overseas and (2) to expand efforts to collect and disseminate foreign technical information to American companies as a means of stimulating industrial innovation.[22] One of the key recommendations was to reconstitute the system of commercial and scientific attachés in American embassies around the world to more closely fulfill these objectives. Subsequently, in 1979 the Commerce Technical Advisory Board formed a subgroup dealing with international issues of scientific and technical information. Representatives from government, industry, and academia will look at subjects such as collecting technical information from foreign countries.

Major U.S. corporations have of course been engaged in keeping up with foreign technological and economic developments not only to help facilitate their own innovative process but also to gain commercial intelligence on competitive products and market trends. Indicative of increased interest in foreign technical information is the subscription service offered by GE highlighting new innovations and technological developments from foreign countries. Multinational corporations (MNCs) in particular must necessarily be sensitive to developments in foreign countries if for no other reason than to protect their overseas investments and market activities. They are also well positioned to monitor foreign technological activity, although the evidence to date points to MNCs more as instruments of technology transfers from the United States. The smaller companies are less attuned to foreign developments and require more governmental assistance because of their more limited resources and their narrower markets. However, the rising tide of foreign competition in the domestic market is a potent reminder to businessmen of every ilk of the growing foreign challenge. So too is the quickening pace of foreign investments in America as more and more foreign enterprises enter the American market by buying up American firms.

Foreign investments in the United States have grown rapidly as foreign firms rush to enter the American market at a time when the U.S. dollar has declined dramatically in relation to foreign currencies. In early 1979, the 100 largest foreign investments in the United States commanded sales totaling $113 billion, which represents a substantial 40 percent increase over the corresponding 1977 figure.[23] West German companies are dominant in accounting for the largest proportion of U.S. sales with the United Kingdom and Canadian investors not far behind. In terms of investment size, the Netherlands lead with an estimated $9 billion, followed by Great Britain at $7.4 billion and Canada at $6.5 billion.[24] During the 1969–1978 period, foreign direct investments in the United States have grown more rapidly

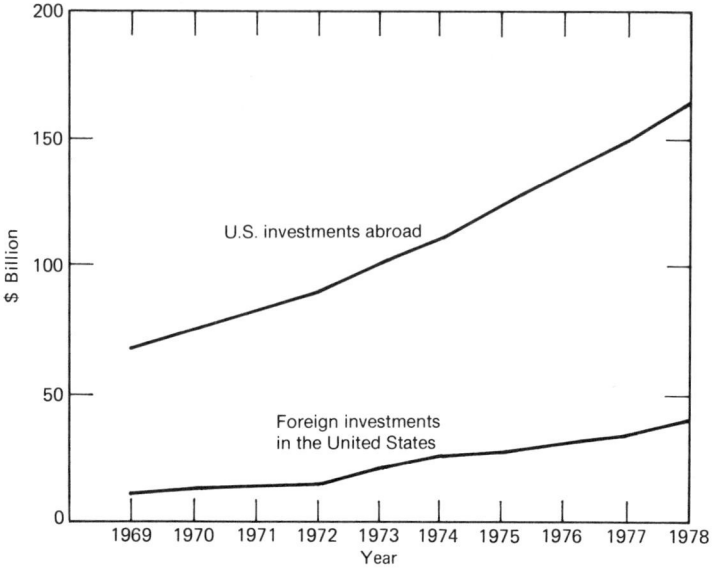

Figure 4-2   Comparison of total U.S. investments in foreign countries and total foreign investments in the United States.

(236 percent) than U.S. investments in foreign countries (141 percent). The total U.S. investment in foreign countries, however, still exceeds foreign investments in the U.S. by a factor of four, as illustrated in Figure 4-2.[25] The rapid infusion of foreign capital in the United States is attracted mainly by comparatively lower costs of production, a cheaper dollar, and the desire of foreign investors to gain and protect their access to the American market— especially in the event import restrictions are imposed by the United States to reduce its trade deficit. The trend thus points unmistakably to a greater foreign presence and competition in domestic markets.

It is not clear that foreign investments in the United States effectively facilitate the import of foreign technology, as market access rather than technology transfer is usually the more important consideration for the foreign companies. There is evidence in some cases that foreign enterprises have transferred technology and know-how to the United States in conjunction with their U.S. investments. On the other hand, the Japanese strategy of establishing commercial operations in America to tap American technological know-how as well as to gain access to the U.S. market would seem to indicate the opposite. Yet it is reasonable to expect that foreign investments in the United States commonly are accompanied by advanced technologies and production know-how if they are to succeed in the highly competitive American market. If American industry was loath to pursue export opportunities aggressively, the same can not be said of foreign industries. As a result, American industry is having to contend with the foreign technological and economic challenge on its own home turf.

Developments in the defense sector have been no less portentous in leading to a more receptive American disposition towards foreign technology. The emphasis on improved NATO cooperation in conventional armaments has produced a much closer involvement in European military and technological affairs on the part of the military services. Efforts to promote a more balanced two-way transatlantic armaments trade necessitates a continuing and comprehensive evaluation of Western European weapon systems and equipment that might appear to satisfy U.S. military requirements. An essential element of the evaluation is to acquire a technological familiarity of the candidate systems as well as a verification of their performance capabilities. Increased NATO armaments cooperation also means developing more efficient use of financial and material resources in allied countries through cooperative development and production of conventional armaments. The selection process to determine European partners in codevelopment and coproduction requires an in depth assessment of the technological capabilities of candidate firms in addition to other factors, such as management record and financial strengths. Opening up U.S. government procurements in general to more foreign suppliers and contractors also requires comprehensive information on the technological capabilities of foreign bidders in addition to management and financial data. Concurrent with efforts to develop greater participation by foreign enterprises in the American defense market is the attempt to establish more effective controls on the export of U.S. technology and conventional armaments. Export license decisions therefore must be based on firm, up-to-date information on the intended recipient country's technological position relative to the United States in addition to other factors, such as the country's reexport policies and geopolitical relationships.

In all these areas knowledge of the state of foreign technological development and ready access to relevant technical information are essential requirements. Recognizing this, each of the military services have initiated modest efforts to systematize the information gathering process. Though each service has proceeded along different paths in this respect, a Defense Department working group was formed to consider the establishment of a consolidated Department of Defense (DOD) corporate memory on foreign technology to serve the needs of the entire defense community. Data on national economies, defense industries, and weapon systems of foreign countries are also considered in conjunction with technological data inasmuch as issues on NATO arms cooperation and export control generally involve the intersection of military, economic, technological, and geopolitical considerations alike. The ultimate objective of course is to avoid duplicative efforts among the services and to provide a useful and timely foreign technology information service for DOD users. The approach taken was to utilize to the fullest possible extent existing data sources world-wide and to consolidate the data in a format that would be readily accessible to the widest DOD user community. The working group activity then represents the first step towards a much needed foreign technology information system that attempts to pull together appropriate data from numerous and diverse primary sources world-wide.

The trend then is unmistakably towards greater U.S. involvement and reliance on foreign markets and technological developments. However, a kaleidoscope of activities concerning foreign technology seems to be materializing as differents parts of government and industry are working with little coordination to satisfy their own informational needs on foreign markets and technology. In one sense, it is a natural outgrowth of America's free-market economy and its decentralized form of government. On the other hand, the lack of centralized planning and coordination produces basic inefficiencies in the overall foreign technology monitoring activity. More importantly, it hampers America's ability to channel information-gathering activities to areas of national need so that what we learn from abroad may more effectively impact on U.S. foreign policy and domestic programs. The effectiveness by which foreign technological information is exploited in the United States is also very much dependent on whether American institutions are properly organized to exploit foreign technology fully in striving to meet national objectives. Institutional realignments will be necessary in all likelihood simply because of America's strong technological self-reliance of the recent past, which must now be tempered with an equally strong respect for foreign technological competence and innovativeness. What these institutional changes must be are yet to be determined as we wrestle with the problem of revitalizing our innovative abilities based on both foreign and domestic technologies. Yet it is encouraging to observe that the increasing awareness of foreign technology has given rise to serious albeit uncoordinated efforts to exploit foreign technological developments without the benefit of White House or Congressional pronouncements on the subject.

Strong government leadership will eventually be needed, however, if the necessary institutional changes are to be made. The United States needs a technology delivery system to translate scientific and technological know-how, domestic and foreign, into competitive goods and services. Enlightened public policies must accompany adequate industrial incentives in order to bridge the technoeconomic gap. However, reviews of U.S. technology policy tend to view the role of the federal government primarily as a resource of technology. It is argued that through decades of R&D funding and internal R&D activities, the federal governmental has developed a sizeable technology base that could be applied to make governmental services at the state and local levels more efficient and could also help stimulate a faster pace of technological innovation in the private sector. However, inadequate attention has been devoted to how the United States might develop increased exploitation of foreign technology. As a result, an important new role for the federal government emerges that requires the government to serve as a link with and a potential user of foreign technology. In this role the government's objective is to smooth the process by which potential users in the public and private sectors could obtain access to foreign technical information. The net result may well help reinvigorate America's ability to innovate.

In the same way that the predominantly outward flow of U.S. technology in the past has helped improve the economy and technological competitiveness of recipient countries, the development of a technology link feeding back into the United States could help to upgrade U.S. capabilities and competitiveness. An improvement in American capabilities could in turn precipitate an increased movement of technology back overseas, thus further upgrading foreign capabilities and stimulating the return flow of technology to the United States, and so on. A mutually interactive technological relationship of this nature will probably be a prerequisite for greater technological and economic progress in tomorrow's more interdependent world. Finally, the extent to which America can adapt to the changing world conditions will determine in large measure the nation's future technological and economic competitiveness.

CHAPTER 5

The Climate for Innovation In Leading Western Industrial Nations

A strong case exists for the United States to devote greater energy towards learning about foreign technological developments and acquiring appropriate technologies to expedite American innovations and to help revitalize the national economy. This proposition represents a tangible and feasible solution to the widely proclaimed predicament in which America finds itself—a stagnating economy against a backdrop of rising foreign technological and trade competitiveness. While increased attention to and use of foreign technology may be an answer, it is by no means sufficient in itself. Complementary efforts must also be undertaken to strengthen the technoeconomic interface in the United States. That is, additional measures need to be instituted by which the nation's technological assets might be better utilized to stimulate national economic growth. Attentiveness to foreign technological developments promises to enhance America's ability to innovate. However, other complementary, potentially effective measures must also be considered. For this reason it is prudent not to focus solely on foreign technological developments but to become attentive to the institutional mechanisms employed in other countries and the experiences of these countries in building a strong technoeconomic interface. The intent, of course, is to attempt to identify appropriate lessons for adaptation to the American scene. It is therefore desirable to select for further assessment two of the leading economies in the Western industrialized world—the Federal Republic of Germany and Japan. These two nations experienced phenomenal industrial development and eco-

129

nomic growth during the post World War II period. The import of foreign technologies in fact played an important part in their economic reconstruction and growth. Moreover, German and Japanese manufacturers have become highly skilled in applying technological advances to the commercial marketplace—particularly using technology developed in the United States. In examining the German and Japanese experience, then, we should attempt to identify those measures employed in the two countries that appear to share commonality as these measures tend to transcend cultural and societal differences and are more likely to be the seminal contributors to their economic success.

Federal Republic of Germany

Since its official birth in 1949, the Federal Republic of Germany—more commonly known as West Germany—has in the space of a few scant decades become one of the most prosperous industrial nations in the world. The nation enjoys high employment, satisfactory growth, slow inflation, a strong national currency, and a large reserve of foreign exchange earnings accumulated from an export-oriented economy. West German workers are among the best paid of any major industrial nation, and they enjoy one of the highest standards of living in the world. Except to buy a single-family home, West Germans pay less for consumer goods than do the citizens of most major industrial countries. Only Americans pay less for consumer goods, clothing, food, and other modern conveniences than do Germans. German exports are respected throughout the world for their high quality and technical sophistication. German exports range from technical know-how to complete steel, chemical, synthetics, electronics, and nuclear power plants, many of which are in direct competition with the United States. The German armaments industry has also been reestablished to the point that a major weapon system such as the Roland surface-to-air missile is being coproduced and deployed as part of the U.S. Army weapons inventory. In addition, the German Leopard tank is considered among the best in the NATO arsenal. West Germany, situated on the front lines of any future conflict with the Warsaw Pact countries, nevertheless manages to maintain an active trade with the communist countries. It is in fact the Soviet bloc's biggest and most important trading partner among the Free World industrial nations. Along with its resurgent economic power, West Germany is emerging as a major international power and is assuming an increasingly strong leadership role in world affairs.

The roots of the German economic miracle trace back to 1948 when Ludwig Erhard's ideas for a social market economy were introduced in place of the controlled economy of the Nazi era. The law of supply and demand thus replaced the artificial controls introduced by the Nazis and subsequently kept in force during the Allied occupation. From this important milestone has sprung the modern industrial power that is West Germany today. Germany's rapid industrial development has occurred despite a history of scarce energy resources. Indeed, the country's lack of energy resources has spawned a sizeable industry that is a world leader in energy alternatives such as coal gasification, gas liquefaction, oil shale, and tar-sand processing.[1] Furthermore, a high energy-conservation ethic, which limits energy consumption to almost half of the U.S. consumption per unit GNP,[2] has not adversely affected the nation's rapid industrial development or its standard of living, which is comparable to that in the United States as measured by income per capita and GNP per capita.

Much of the credit for Germany's industrial progress is attributed to the spirit of cooperation existing between management and labor. A policy of consensus and responsible behavior rather than of confrontation marks relations between the two "social partners." The trade unions rarely go on strike but generally work with management to determine equitable wage increases consistent with expected product price increases. The manner in which labor-management negotiations are conducted is notably different from the way such negotiations are conducted in America. American unions and companies bargain as autonomous organizations and do not belong to strong, central federations. In Germany there are only 16 industry-wide unions, and these are organized under the German Trade Union Federation. Company managements are also organized and represented by the Federation of Employers' Association. Collective bargaining is conducted between these two groups for entire industries and branches of the economy instead of for each individual company and trade union. The consensus reached at the bargaining table then applies to all workers, union and nonunion, in that particular industry. This form of collective bargaining has no parallel in the United States.

Another unusual aspect of German labor-management relations is the nearly equal voice workers have in the management of their company. In every company, large and small alike, worker-delegates are elected to represent the interests of labor to owners and managers. In large corporations of 2000 people or more, worker-delegates comprise half the members of the corporate board of directors. In this way German labor is very much integrated into the mainstream of economic life. Instead of confining union interests exclusively to wages and working conditions as in America, the workers get involved with the management of corporate assets and consequently become closely familiar with problems of management and production. Their more intimate understanding of management problems and issues makes it easier for labor-management consensus to be reached.

Close identification with company management provides added fuel for the German workers' penchant for avoiding waste and strife. The efficiency and productiveness of the German worker are well known. In addition, German workers generally are not spendthrifts but tend to save their money. The German savings rate amounts to about 14 percent of disposable income—roughly triple the American rate. A major factor is the widespread recognition that, despite recent prosperity, the German workers want to build a cushion to soften possible future periods of high inflation or recession. The tendency towards consumer savings translates into greater availability of investment capital for industrial development, which in turn has helped the nation achieve one of the fastest productivity growth rates among the Western industrial nations. These worker attributes coupled with responsible and moderate worker wage demands are major factors West Germany's ability to control inflation and to compete effectively in world markets in spite of the comparatively high labor wage-rates prevailing in the country.

The close relationship of the trade unions to German business is amply illustrated by the Trade Union Federation, which is itself a big business. The unions in fact own a wide range of businesses such as newspapers, insurance companies, manufacturing concerns, one of the nation's largest banks, plus the nation's largest real estate and homebuilding company. The union business enterprises provide employment for 25,000 people and therefore is hardly an insignificant factor in the economy.

The unusual extent of labor participation in business and corporate management portrays to some observers a creeping socialism that is slowly enveloping the market economy in Germany.[3] That is, Ludwig Erhard's social market economy is being slowly transformed into a socialistic economy. However, the evidence to date seems to indicate that these concerns are unfounded. There appears to be a strong German consensus on the superiority of free enterprise over socialism. In the mid 1960s, the Social Democratic party in Germany dropped nationalism as a political issue and subsequently was elected to power in 1969. In the intervening years, especially during the world recession of 1974–1975, the Social Democrats acted very much like a free enterprise government. Government did not intervene to help weak enterprises. There was no attempt to bail out weak enterprises in order to preserve jobs as was the case in Great Britain, France, Italy, and even in the United States. Efficiency was not penalized in order to subsidize inefficiency. A well known example is Volkswagen, which was left to weather on its own the downward business trend brought on at the time by spiraling labor costs and the appreciation of the German Deutschmark. Evidence of the free enterprise spirit such as this undoubtedly has helped keep the Social Democrats from having to relinquish the reins of power.

The central government then generally favors a laissez-faire rather than an interventionist or restrictive industrial policy. Although the government encourages entrepreneurship and innovation through a host of indirect measures, it nevertheless limits its direct support to a few high-technology program areas identified to be in the national interest.

The priority program areas are nuclear energy, space research, data processing, civil aviation, and oceanography.[4] In nuclear energy Germany is now competitive in the export market. In space research Germany cooperates with other nations to develop and launch earth satellites. In civil aviation the focus of attention has been on the European Airbus consortium. In data processing the emphasis has been in development of minicomputers, peripheral equipment, data processing applications, and training facilities. In oceanography German efforts have centered on mining the world's oceans for food and mineral resources. Projects in these areas are initiated by either government or industry. Both research and industrial application activities are pursued. Industrial incentive is maintained by requiring industry to provide a portion of the funding. Technology developed from the projects are owned by the participating firm, but the government may make the technology available to other firms in which case negotiations are conducted directly between firms.

Government support for science and technology is channeled through a labyrinthine conduit. Government responsibility for science is shared between the federal and state governments. Plans, policies, and priorities are prepared in conjunction with universities and industry. Recommendations are made to the federal government from numerous coordinating bodies, advisory councils, and evaluation committees concerning the establishment and implementation of science policy. Funding for the priority programs of national importance comes from the Federal Ministry for Education and Science [Bundesministerium für Bildung und Wissenschaft (BMBW)]. The funds are made available generally to nonprofit or limited-liability companies formed in cooperation with one or more of the states to pursue research in the priority program areas. The BMBW is also the federal funding agency for the privately operated but publicly supported, nonprofit organizations that implement government science policy. The government does not maintain its own research laboratories but relies on these nonprofit organizations. The key organizations in this category are the German Research Society [Deutsche Forschungsgemeinschaft (DFG)], the Max Planck Society for Advancement of the Sciences [Max-Planck-Gesellschaft zur Förderung der Wissenschaften (MPG)], the Fraunhofer Society for the Advancement of Applied Research [Fraunhofer-Gesellschaft zur Förderung der angewandten Forschung (FhG)], and the Confederation of Industrial Research Associations [Arbeitsgemeinschaft Industrieller Forschungsvereinigungen (AIF)].

The DFG is mainly concerned with promoting basic research, primarily in universities but also in other supraregional research associations and societies. It does not operate its own research laboratories but supplies the funds for the support of existing research facilities or for the formation of new ones. Most of the funds available to the DFG come from the state governments and from the federal government via the BMBW. The DFG attempts to develop more efficient use of research facilities and resources by concentrating specific areas of research in the best qualified university research institute and preventing duplication of work at other universities. It also works to promote cooperation between universities and other research centers.

The largest research establishment in Germany is the MPG with its 52 research institutes and centers. Like the DFG, funding comes mostly from the federal government via BMBW and the state governments. Research associations and industrial contracts account for the remaining sources of funds. Although the MPG too conducts mostly basic research, it is also involved in transferring the research results to industrial applications. The primary link to industry is provided by the Garching Instrument Company for the Industrial Use of Research Results, a for-profit organization that serves to bridge the gap between basic research and commercial applications. Garching keeps abreast of new technical developments in the MPG institutes as well as the market needs in industry, and it generally serves to close the information gap between research and the market. It also negotiates agreements with industry, acts as licensor for MPG patents, engages in prototype development, and in some cases its activities extend to production and product sales. Garching thus acts to commercialize technical developments of the MPG institutes.

The FhG has prime responsibility for performing applied research in contrast to the basic research activities of the MPG. Research results are published, but in the case of proprietary research, publication may be delayed up to two years. Like the MPG, the FhG operates its own research institutes and is deeply involved with transferring its research results to industry. Most of the 20 FhG institutes have their own boards of directors composed of representatives of state governments, universities, and industry in roughly equal numbers. Funds for the FhG institutes are provided by BMBW as well as from governmental and industrial research contracts. The FhG also works to improve the transfer of technology among its FhG clients—between industries and between the government and industry. One of the FhG institutes is the Institute for Technology and Innovation Research, which engages in studies relating to technology transfer and industrial innovation. The activities of the FhG institutes are meant to benefit the entire industrial sector but particular emphasis is given to small- and medium-sized firms with no R&D capabilities of their own. The FhG with federal and state government funds also operates the Patent Bureau for German Research which provides assistance to independent inventors in gaining protection for their inventions.* The Patent Bureau examines and tests inventions for economic promise without charge, provides loans to cover patent application fees, and acts as mediator between inventors and potential industrial licensees.[5] The activities of the Patent Bureau, like those of the Garching Instrument Company, have no corresponding equivalence in the United States. Another group under the FhG is Arbeitsgruppe für Patentverwertung (ARPAT), which basically serves as a clearing house of patent rights arising from publicly supported R&D. ARPAT acts as a licensing and patent rights broker but does not participate in negotiations between the principal parties. If necessary ARPAT mediates on license negotiations. If negotiations fail, the federal government has the authority to grant licenses to interested third parties even against the will of the patent-right holders.

The AIF is the principal organization for industrial research in Germany. Its aim is to promote cooperative research among the 76 industrial research associations comprising its membership. The research associations them-

---

*Free advisory services are also provided to independent inventors with limited financial means by the cities of Stuttgart and Ludwigshafen.

selves were formed by their member companies to pool R&D resources in their respective industrial sectors. The member firms are mostly smaller companies with limited or no R&D facilities. The actual research work is conducted in approximately 150 industrial research institutes established by the respective industrial sectors either at universities or as independent organizations. The research associations in AIF are representative of all sectors of German industry except the chemical and electrotechnical sectors. Companies in these two sectors necessarily have developed their own R&D facilities owing to the high technological content of their products. The AIF coordinates the research activities of the research associations and represents them in dealings with the federal government. The major portion of funds for AIF is derived from industrial contributions and membership fees. The remainder of its funding support comes from the federal government in the form of grants. The government does not expect repayment of the public funds but anticipates economic returns to come in the form of higher industrial productivity, profitability, and consequently higher corporate tax revenues.

The existence of numerous publicly supported research institutions is in a major sense a reflection of the industrial structure present in Germany. Industry is dominated by a small number of large companies that perform the bulk of government-supported R&D. These large firms are well-equipped and have the technical capability to provide competent R&D services. Small companies, on the other hand, generally lack the R&D facilities and consequently must rely on appropriate research institutions in their particular industrial sector. The research institutions therefore represent an aggregration of R&D resources serving the needs of the smaller firms. Consequently, government funds are made available to a smaller population of qualified big companies, universities, and research institutions acting as proxy for smaller firms. The size of the government R&D market in any event is insufficient to support all the small firms desiring government R&D business. By thus concentrating limited government resources in the hands of able performers representative of the breadth of industry, the federal government attempts to diffuse the results of selective government R&D efforts to the widest reaches of industry. Another important consideration is that government attempts to influence the direction of industrial research to help meet national goals becomes a much more manageable task when there are fewer R&D performers.

It is readily evident from the emphasis given to industrial applications by the federal government and research institutions that great importance is assigned to developing an industrially relevant national R&D program. Determined efforts are made by the government to generate economic benefits from the national investment in science and technology. In order to obtain government financial support, however, industrial firms must be willing to share project costs. This cost-share arrangement is as much a gauge of a firm's earnestness and conviction of a project's industrial applicability and worth as it is a factor in fostering a genuine spirit of government-industry cooperation. The government encourages the retention of technological advances derived from the project to be exploited in the industrial context. In most cases, however, the results must be published in the open literature to promote further the diffusion of the new development throughout industry. Often this requirement is a source of contention between the government and the company when proprietary knowledge is involved. Nevertheless, the general government policy is to encourage the unrestricted flow of industrial technology between industrial sectors and also with foreign firms.

Indeed, technology exports are encouraged as much as imports of technology no matter whether through the purchase or sale of licenses, direct investments, or other means. An indication of Germany's unrestrictive technology transfer policy was found earlier (Table 3-3) in analyzing the source of inventions underlying the 500 significant innovations identified during the 1953–1973 period. Further evidence of Germany's liberal technology transfer outlook is found in the nation's sizeable trade with communist countries. Germany is the biggest and most important capitalist trading partner for the Soviet-bloc countries and the Peoples' Republic of China. Germany's liberal technology export policy has in fact evoked some concern on the part of industry that perhaps greater controls should be imposed on the sale of technology as compared to sales of manufactured goods. It is interesting to contrast this situation with the prevailing perspective of a large part of American industry that U.S. government controls on technology exports are so restrictive that they impede unduly the U.S. competitive position in world markets.

The federal government has devised further measures to stimulate industrial innovation. In 1971 it enacted a measure to authorize public grants to assist companies in introducing new innovations into the commercial market. The innovation may be either a product or process, but it must be technologically new and have good commercial promise. The objective of this measure is to accelerate the first market introduction of significant new innovations that otherwise would not be commercialized or that would be possible only after considerable delay. The grant-funds under certain circumstances are repayable to the government after the commercial success of the innovation. The government has also moved to make risk capital more easily available to small- and medium-sized technology companies. One approach is to provide credits and guarantees to equity investment companies in order to encourage them to invest in technological-innovation projects of the small- and medium-sized technology firms. Another measure is to share risks with a private venture capital company for a limited period of time. The government will share the possible losses of a private venture capital company, founded for this particular purpose in 1975, up to an amount equal to the paid-in capital of the company not exceeding DM 50 million. Government funds are repayable with modest interest after successful completion of the specific project venture. The rationale behind this measure, of course, is that with government risk sharing, venture capital will be more easily available to the small- and medium-sized firm for technological-innovation projects.

The government's role in promoting technological innovation is by no means limited to direct-funding schemes but also extends to more indirect measures that influence the general economic climate, affecting the individual inventor, entrepreneur and the established technology-oriented firms. The general policy is to produce a favorable climate for industrial innovation by means of appropriate government measures. Central to the federal government's efforts to stimulate a favorable environment for innovation is the provision for unrestrained competition in the marketplace. In 1958 the Law Prohibiting Restraint of Competition was enacted in order to stimulate competition and to remove restraints on a free-market economy. This law, since amended in 1966 and 1973, lays the foundation for the nation's free-market economy and is the major stimulus for a favorable climate for industrial innovation. In addition, the government acts to assist industry in conducting market research by publishing pertinent information on the economy, for example, the aims of the nation's economic policy, medium-term prospects for economic development, and economic reports and forecasts on different industrial sectors. Recognizing that a free-market economy is a dynamic and continually evolving structure where factors of production such as labor and capital must necessarily be shifted continually to respond to changing market conditions, the federal government also instituted several provisions aimed at encouraging the mobility of people and capital in the economic sector. It provides free job placement and advice, provides funds to encourage personnel advanced training or retraining, makes grants to employers to create jobs, and also helps cover costs of employee relocations.

Additional indirect measures have been enacted aimed more specifically at developing greater entrepreneurial incentives among the scientific and engineering community. These measures fall into two major categories— legal and fiscal incentives. In the first category is the German patent system and associated legislation; the second category consists of a variety of tax benefits such as credits, allowances, and accelerated depreciation for R&D investments and expenditures.

The German patent laws are basically designed to promote the introduction of new innovations into industry, to protect the rights of inventors, and to provide adequate incentives and rewards as an inducement for inventiveness and creativity. Speeding the introduction of new innovations into industry is accomplished in several ways. Upon application of a patent, the inventor may choose to defer examination of his invention for patentability and commercial usefulness for up to seven years. His legal title to the invention is established by the filing date of the application. Because there is a nominal fee for requesting a full examination, the inventor generally weighs his decision carefully before proceeding. Meanwhile the German Patent Office is not overburdened with a high examination case load and can therefore prosecute the fewer examination requests it receives more rapidly. To spur the diffusion of the invention technology to potential industrial users, however, all patent applications as a rule are made public 18 months after the filing date. After a patent is awarded, the inventor enjoys legal protection on his invention for 18 years. Because the maintenance fees are a significant expense and rise yearly, the inventor normally tries to commercialize his invention as quickly as possible. He also has recourse to reduce fee payments by 50 percent merely by declaring when filing his patent application that his patent would be available for licensing. This provision, of course, is intended to help spur the commercialization process. A further inducement for rapid application of the invention stems from the availability of patents to be used in research work without requiring a license. Only after a new application develops, does the researcher need to negotiate with the inventor for a license.

The rights of inventors in Germany are closely guarded and include substantive and tangible incentives and rewards for his efforts. In the case of independent inventors, their income from activities outside of a position of employment as well as his expenditures relating to an invention qualify for special tax treatment. For example, income from an invention accrued not as part of the inventor's own business is subject to reduced taxes. Additionally, capitalized expenditures on inventions that are used in the inventor's own business may be fully depreciated. On inventions that are job related, the employer must lay claim to the invention within a given period of time. Otherwise, the rights to the invention reverts to the employee. Claims to the invention may be either unlimited or limited.[5,6] Unlimited claims are those where the employer retains all rights to the invention while the employee retains only the right to be named the inventor. On limited claims the employee holds all rights to his invention but the employer has the right to use it. For every invention reported the employee-inventor is entitled to just compensation from this employer. The employee may even be compensated for nonpatentable technical improvements from which the employer derives commercial advantages. Furthermore, the employee can be compensated for suggested improvements that neither qualify as an invention nor a technical improvement, as long as the employer utilizes them significantly. In cases where conflict arises between the employer and employee, an arbitration board set up under the German Patent Bureau mediates the dispute to arrive at an amicable settlement without having to resort to legal proceedings.

In addition to these favorable provisions of the patent laws, a broad array of fiscal incentives in the form of tax benefits and government subsidies have been instituted to encourage scientific activity and capital investment by industrial enterprises. Individual taxpayers who are wage earners or self-employed can request tax relief on supplementary income derived from scientific activity. Supplementary income up to 50 percent of the taxpayer's regular income from wages or self-employment is eligible for the tax relief. All expenditures made for scientific purposes are also tax deductible in the year incurred. Corporations that donate real property to scientific or educational institutions may write off such property gifts for tax purposes. The gifts moreover are tax free to the recipient institutions. Also, public grants for scientific activity are tax free for the recipient organization. In the case of assets used for R&D, special depreciation rates of up to 50 percent for movable goods and 30 percent for immovable items (e.g., buildings) may be allowed in addition to normal linear depreciation for wear and tear. This additional depreciation is allowable during the year of acquisition or construction of the asset plus the following four years. The asset must be used for R&D purposes for at least three years following its acquisition or construction. Movable assets must be used exclusively for R&D while at least two thirds of the immovable asset must be used for this purpose during this period. As an additional inducement for corporate R&D investment, another 10 percent deduction is allowable if the asset is used for basic research, new methods or products, or the improvement of new methods or products of significant importance. On top of these tax incentives, firms that invest in industrial R&D facilities are also eligible for a 10 percent government subsidy from the BMBW.

Thus there is an unmistakable emphasis on industrial innovation that undergirds government-industry relations in West Germany. Through a broad system of direct-funding support and indirect policy measures aimed at providing adequate legal and fiscal incentives to industry, the federal government acts to stimulate the climate for innovation as well as to provide tangible incentives to industry at all stages of the technological innovation process.

Japan

Japan is a land with practically no natural resources and an agricultural output insufficient to feed its population of more than 100 million people. However, the land is not lacking in a nation's most important resource—a well-educated and industrious people. This vital resource has in the space of two generations transformed Japan into a leading industrial power rivaled only by the United States. The innovativeness of the Japanese people has enabled the nation to adjust well to its lack of natural resources. It has provided for the creation of efficient production processes that conserve raw materials, energy, and physical space. It has also produced lower cost, more competitive goods, and higher value-added export products that contribute to a favorable trade balance and pays for the nation's sizable raw-material imports.[7] Since World War II, Japanese ingenuity and industriousness have powered the nation's economic engine to new heights. The national GNP between 1950–1970 grew by an average annual rate of 10 percent, which is double the growth rate of European industrial countries.[5] Although its per capita income trails that of the United States, West Germany, and France, its inflation rate has been among the lowest of the Western industrial nations. While the United States, France, and Great Britain were battling double-digit inflation late in the 1970s, Japan and West Germany were keeping their inflation rates under control in the 4 percent range. In foreign trade Japan has been steadily increasing its share of the world market, realizing high trade surpluses, especially in trade with the United States. Moreover, the nation is effectively challenging America in world markets for technology-intensive products such as computers and semiconductor electronics.

How did Japan succeed in progressing so rapidly? Several prominent reasons are often cited:[8] a close cooperation between government and industry, a diligent and company-loyal labor force, an emphasis on company growth instead of profitability, a management-by-consensus philosophy that is people centered rather than profit centered, a close attention to productivity improvement, and a disposition to profit from foreign technology. All these factors undoubtedly played a part. Perhaps the most important factor is the Japanese knack for technological innovation and their willingness to take risks associated with entrepreneurial undertakings. The reason this is so important is that in trying to capitalize fully on its most important national resource—the knowledge and skills of its people—Japan recognizes that its future lies in the direction of knowledge-intensive industries that make full use of the innovative ability of its people.

A major source of industry's investment capital is derived from consumer savings, which typically are close to 20 percent of disposable income.[9] This exceptionally high rate reflects the frugal nature of the Japanese people. In addition, savings are encouraged by a preferential tax rate on savings deposits, an inadequate social security system, and limited use of installment credit in consumer purchases. Personal savings deposits in the banks are an important source of business loans to industry and have helped fuel much of Japan's industrial growth. Industrial investments are financed mainly through extensive bank credits. Japanese companies typically assume debt-to-equity ratios of 4:1 or more. This helps explain why industry is oriented more towards growth rather than profitability as meeting interest payments on its high debt structure becomes a higher priority than dividend payments to stockholders. The high debt-structure of Japanese industry is also indicative of the pivotal importance of Japanese banks in furthering industrial development. However, because capital requirements in the rapidly expanding economy often outstrip resources of the commercial banks, government financial institutions also play an important role in making investment loans regularly. Japan's financial institutions in turn are closely regulated by the government. In this manner the Japanese government greatly influences the directions of industrial growth. Capital resources as a result tend to be concentrated in the industrial areas that hold the brightest promise of market growth; declining industries, on the other hand, are assisted in reallocating resources to more fruitful fields or are allowed to die when no longer competitive.[10]

Japanese banks, therefore, play a pivotal role in the exceptionally close relationship between government and industry. The highly leveraged Japanese industry is heavily dependent on the banks for loans on favorable terms for both working capital and investment funds. Government, on the other hand, guides and influences the banks to finance industrial projects and capital investments that have high priority and are in accord with national goals and plans. The banking institutions are therefore indispensable in bridging the interface between government and the private sector. However, the close relationship between government and industry does not revolve solely around the financial sector. Direct communications also exist between government and industry at all levels. Mutual decisions are made regarding economic policies and planning. New programs are undertaken by the government only after prior collaboration and consensus has been reached with industry. Industrial views are represented on numerous government advisory councils and committees, allowing industry to be involved in the government planning process at the very early stages. Moreover, industrial managers and business executives maintain close contacts with government officials. Many of them in fact previously worked for the government until reaching the mandatory retirement age of 55. Civil servants are also normally rotated among several different government agencies during their careers so that they possess a breadth of knowledge on the inner workings of government. Not all industrial sectors uniformly enjoy close working relations with the government, however, as government attention is primarily devoted to those sectors considered high priority for attaining the nation's economic goals.

The character of Japan's industrial structure is one that has reached a high state of industrialization in the heavy and chemical industries (e.g., primary metals, metals fabrication, machinery, chemicals, and transportation and electrical equipment) but where increasing emphasis is being shifted to knowledge-intensive industries such as computers, telecommunications, and nuclear energy.[5] In the process, Japanese industries have become pollution generating and highly energy consuming, prompting new investments to be made in the fields of environmental protection and alternative energy sources.

National priorities have been established in energy, space, computers, and ocean resources. Direct government support is provided in the form of fully subsidized projects that make up the National Research and Development Programs (NRDP) established in 1966. Some of the development projects concern magnetohydrodynamic generators, super high performance computers, salt-water desalination, and remote-controlled undersea oil-drilling rigs. NRDP projects such as these are aimed at developing whole new industries that will create competitive opportunities abroad and improve the quality of life at home. In addition to the fully subsidized NRDP projects, joint ventures are also undertaken with industry in three major areas: nuclear energy, space exploitation, and ocean development. Industrial participation in the joint projects may range anywhere from 3 to 50 percent. Industrial participants furthermore share in the commercial realization of the project in relation to its share participation in the project. Clearly, these priority projects are all in technology-intensive fields requiring sizeable investments in R&D and are in consonance with Japan's goal of promoting knowledge-intensive industries.

That national science and technology policy will necessarily be inextricably interwoven with economic and industrial policies was in fact recognized by Japanese planners soon after World War II. Indeed, perhaps the most distinguishing feature of Japan's technology policy is its close identification with economic growth. However, industrial reconstruction took precedence immediately after the war, and Japan's policy concerning science and technology was to continue the time-honored principle of importing foreign technologies for the reason that they provided the most rapid and least costly method to catch up with the industrialized West. The R&D situation in Japan during the 1950s therefore was very much like that of a developing country, where the emphasis is on importing foreign technologies and strengthening domestic R&D capabilities to ease the technology assimilation process. Imported technologies and foreign investments were closely regulated by the Japanese government as a result of the Foreign Investment Act enacted in 1950. Under this law the type and flow of technologies into Japan were controlled in order to facilitate national industrial planning and to provide incentive and protection for infant industries against strong international competition. Subsequently, as industries matured and became better able to compete in world markets, most types of foreign investments and technology imports were decontrolled by 1968 although restrictions still exist as to the extent of foreign holdings permissible in Japanese firms.

The 1960s also saw a marked change in Japan's technological policy from one of foreign dependence to one of increasing self-reliance. It was recognized that Japan's heavy dependence on foreign technologies—most of which originated in the United States—was growing too costly and jeopardizing the nation's technical and economic independence.[5] During 1960–1965, fully 25 percent of industrial R&D expenditures constituted payments for foreign licenses. As a result, the government adopted a policy of encouraging independent R&D for developing indigenous technologies and innovations. The approach was to strengthen the nation's technological infrastructure, raise the quality of research, and improve the climate for independent R&D. R&D expenditures were increased many fold, and big science projects were started such as those of the NRDP and large government-industry joint projects such as in space exploitation and ocean development. Direct support measures such as these were further augmented with indirect measures such as R&D tax incentives introduced in 1967 followed by institutional initiatives to expedite the flow and use of technical information. The results of these measures to stimulate indigenous technologies and capabilities are reflected in Japan's record of payments for foreign technologies and receipts from sales of Japanese technologies overseas. Between 1960 and 1967, Japanese payments for foreign technologies climbed from $95 million to nearly $239 million, while receipts from sales of Japanese technologies climbed considerably more rapidly from $2.3 million to $26 million during the same period.

The 1970s witnessed a further adjustment in Japan's science and technology policy as government priorities for environmental protection and social betterment were introduced along with established priorities for economic growth and national technological progress. Japan's preoccupation with economic growth was reassessed in light of adverse social consequences arising from the nation's rapid postwar industrial boom. Quality-of-life issues concerning pollution control, energy conservation, housing, and health came to the fore. It was during this period that oil prices ballooned and that the Sunshine project was initiated to seek alternative energy sources and to reduce Japan's high dependence on foreign energy imports. The nation's technological resources were employed to improve the quality of life of Japanese citizens as well as to fuel further industrial progress. Japan's science and technology policy consequently had to be modified to accommodate both social as well as economic goals. Societal considerations are readily identified in Figure 5-1, which is a schematic of Japan's science and technology policy. As is clear from the figure, technology transfer and technology assessment, that is, technology's impact on society, are integral elements of this policy.

Science and technology policy formulation and implementation involves a number of government agencies, advisory councils, quasi-public corporations, and R&D laboratories. Every government ministry is involved in science and technology to some degree, but the key organizations with specific mandates are shown in Figure 5-2. Overall responsibility for science and technology resides in the Office of the Prime Minister. Matters of policy are the purview of the Council of Science and Technology (CST) and the Science Council of Japan (SCJ). At the operational level are the Science Technology Agency (STA), the Ministry for International Trade and Industry (MITI), and the Ministry of Finance (MOF). Within MITI are several im-

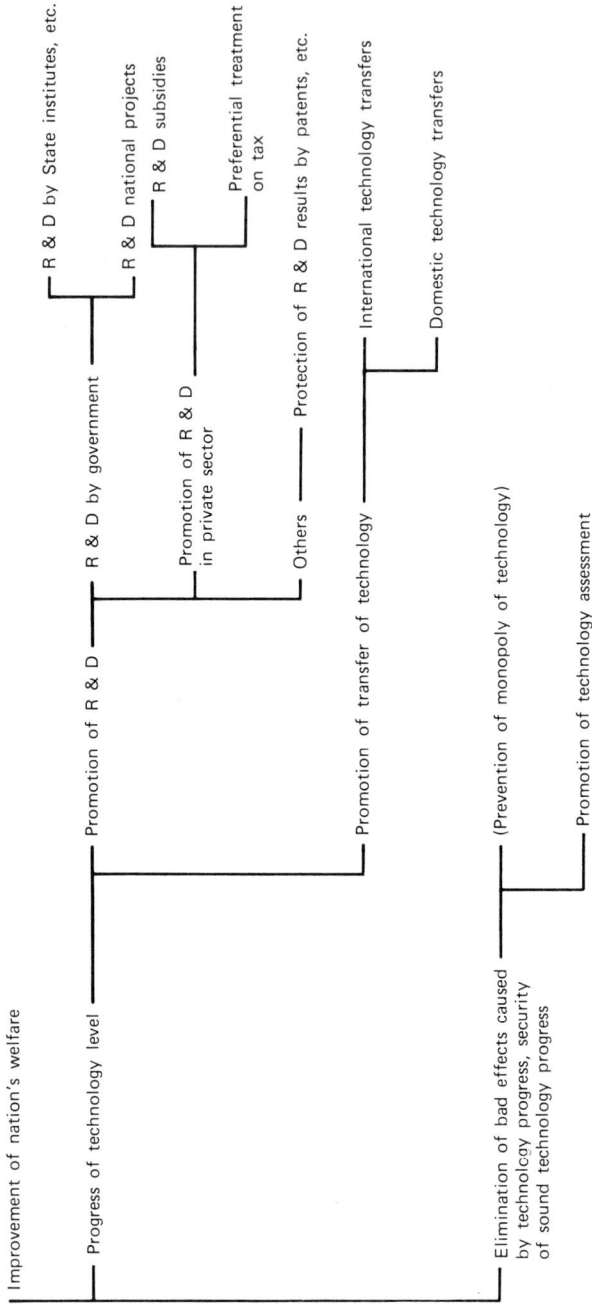

Figure 5-1 *Science and technology policy in Japan. (Source: Reference 5-4)*

portant components: the Agency of Industrial Science and Technology (AIST), the Agency for Small and Medium Enterprises (ASME), and the Patent Agency. The Research Development Corporation of Japan (JRDC) is one of the specially chartered quasi-government corporations administered by the STA. Other organizations of special note are the Japanese Development Bank (JDB), the Japan Industrial Technology Association (JITA), the Japan Patent Information Center (JAPATIC), the Industrial Technology Promotion Association (ITPA), and the Keidanren or the Federation of Economic Organizations. A brief review of the charter and functions of these organizations will help develop further insight into the Japanese R&D infrastructure.

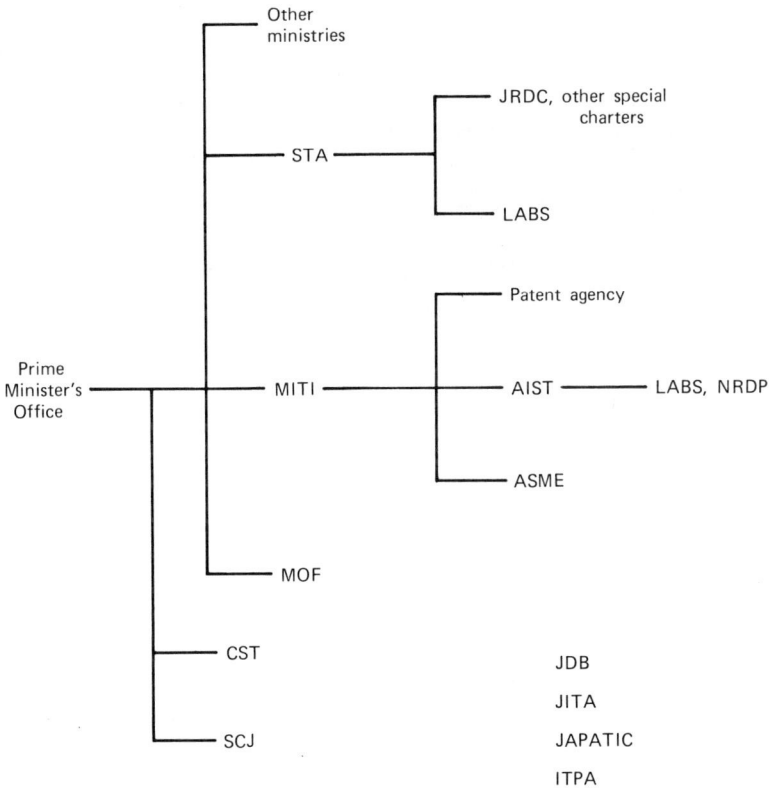

Figure 5-2 *Key government institutions concerned with science and technology in Japan.*

The CST is a cabinet-level science and technology advisory group chaired by the Prime Minister. It is composed of four cabinet members, five appointees, the President of the SCJ, plus the Prime Minister. The function of the CST is to conduct long-range science and technology policy planning and to review research projects in coordination with the SCJ. The SCJ deals in turn with science, the utilization of research, scientific education, and relations with industry. Both councils make recommendations to the Prime Minister. Policy decisions are then developed and implemented by the STA and MITI.

The STA was set up in 1956 to develop and carry out science and technology policy decisions. A member of the cabinet, the Minister of State for Science and Technology, heads the agency. The major functions of the agency are to (1) formulate and implement science and technology policies based on directives from the Prime Minister, (2) coordinate research activities among the different government ministries, (3) participate in formulating the R&D budgets of the different ministries, and (4) administer six government laboratories, seven special charter quasi-government corporations (among which is the JRDC), and four operational bureaus engaged in atomic energy development and other interdisciplinary projects. The STA also serves as the secretariat for the CST and other advisory groups to the Prime Minister and acts in a liaison capacity between the SCJ and the rest of government.

Under the STA is the JRDC, a nonprofit, quasi-government, special charter corporation formed in 1961 and patterned after the National Research and Development Corporation (NRDC) in the United Kingdom. Its charter is to encourage innovation by promoting industrial exploitation of research results and inventions. JRDC exercises its mandate in two basic ways. First, it selects high risk research projects having good commercial potential, negotiates a contract for the commercialization effort with the "commissioned" firm and partially underwrites the cost of the project. JRDC typically underwrites 60–80 percent of the project cost, and the amount is considered an interest-free loan repayable to JRDC if the project is a success within five years. Any royalties realized are shared equally between JRDC and the inventor. If the project is judged a failure, the entire investment is written off as a total loss. The project selection criterion is quite stringent, however, giving rise to a success rate of better than 90 percent. The second method of operation for JRDC is as an intermediary or broker between the government inventor or researcher and a typically small company with the capability to commercialize the invention or research result. For this service JRDC receives 10 percent of the royalties, the inventor receives the remaining 90 percent. JRDC obtains approximately two thirds of its operating funds from royalty income and return of investments from commissioned projects. The other one third of its budget comes as government appropriations from STA. Universities, colleges, and government laboratories are the source of most of the projects undertaken by JRDC.

MITI became involved in science and technology through the R&D connection with industrial growth. Its broad mandate is to plan and coordinate the efficient use of domestic and foreign resources for developing Japan's industrial growth and international competitiveness. MITI has therefore sought to regulate the import of foreign technology and capital in order to protect domestic industries. It has guided and influenced the direction of industrial growth into the more potentially promising areas. MITI has also assisted domestic industries by providing analyses of export markets and by helping to identify promising foreign technologies for potential acquisition. The science and technology functions of MITI reside mostly within the AIST and the ASME.

AIST was formed in 1948 to help industrial firms absorb foreign technologies and exploit them commercially. Subsequently, AIST became the arm of MITI responsible for the ministry's activities in industrial science and technology. Its basic mission is to promote industrial R&D to strengthen the mining and manufacturing industries. AIST also promotes industrial standardization for the purpose of improving mining and manufactured products, modernizing production processes, and providing better commercial services to the consumer. Government subsidies for a broad range of R&D projects come through AIST. Projects involve a wide range of advanced technologies aimed at strengthening the competitive position of industry as well as at meeting social needs relating to environmental protection, industrial waste treatment, auto safety, and the like. Some of the projects are performed within the agency's 16 research institutes and laboratories. Other projects performed in industry include those of the NRDP which was originally started by AIST in 1966 and for which it retains management responsibility. AIST also allocates loans to industry to finance the industrialization of new technologies from funds made available from the JDB. The intent of the loans is of course to help overcome the large costs encountered in the final stages of the innovation process where prototype development and manufacturing start-up expenses are typically much higher than R&D costs.

Small- and medium-sized enterprises receive special attention from the ASME within MITI because of the special problems peculiar to their size, such as inadequate technical resources and inability to obtain credit on favorable terms. Moreover, over 99 percent of the private sector in Japan is made up of small- and medium-sized firms having no more than 300 employees. Established in 1950, the ASME provides technical assistance to these smaller firms in the form of consulting, educational, and training services and access to public research institutions in order to supplement their lack of technological resources and to improve management capabilities. The ASME also provides financial assistance as well as tax and other incentives to stimulate plant modernization and resulting higher production efficiencies.

The MOF is the ministry with which both STA and MITI work closely in formulating plans and budgets in support of their respective R&D projects and activities. STA, in formulating the R&D budgets of the other ministries, collaborates with MOF in establishing funding priorities for the many government programs. MOF interacts closely with MITI in establishing funding levels for subsidized R&D projects and in granting fiscal incentives to industrial firms. While MOF acts only in an advisory capacity regarding the allocation of public funds, it has the authority to make decisions on matters relating to fiscal incentives. The MOF must also be consulted before loans involving tax rebates can be approved by the JDB.

The JDB is the key government financial institution for stimulating economic reconstruction and industrial development. Started in 1951 with public funds, the bank is a government agency that nevertheless retains a degree of independence by being able to raise funds through the issuance of foreign currency bonds. The JDB lending operations are intended to support the government's strategies for industrial development. Loans granted by the JDB usually carry terms more favorable than those available at commercial banks. The bank provides developmental financing for the expensive stages of the innovation process downstream of the R&D phase. It is the source of funds for credits granted by AIST to finance high-risk industrial ventures. JDB credits are applied towards helping set up new industries, modernizing existing plant and equipment, and restructuring the industrial economy to be more in line with evolving markets and development strategies. JDB loans are also used as incentives for businesses to utilize domestic technologies. Lately, because of the increased consciousness concerning the social consequences of technology, the JDB during the 1970s has had to include technological assessment in its procedures for evaluating loan applications. Its criterion for evaluating loan applications has necessarily been broadened to include environmental and other social issues in addition to production and market factors.

Another key institution for promoting fuller utilization of technology is the JITA. The JITA was established in 1969 to foster the transfer of technology both domestically and internationally. The association works to facilitate the dissemination and exchange of technical information between the AIST research laboratories and foreign countries. JITA provides an advisory service on patents and works to inform Japanese industry of promising research results of the MITI laboratories. It generally tries to keep abreast of technological developments at home and abroad and acts to facilitate the transfer of technical information and technology between the source and potential user. In this manner it helps to expedite the adaptation and use of technological advances in the industrial sector.

Another organization set up to facilitate the flow of technical information was JAPATIC. This private, nonprofit group was formed in 1971 to serve as a clearing house for patent information of both Japanese and foreign origin. The JAPATIC is supported by contributions from industry and private organizations and enjoys the full cooperation of the Japanese Patent Office within MITI. Complementing the activities of the JAPATIC is the ITPA, which was formed by the private sector to promote commercial exploitation of government-owned patents, especially those belonging to AIST. Promising government patents are selected and made available to ITPA to seek commercial licenses. Nonexclusive licenses are authorized by the government; ITPA receives 10 percent of the royalties for these licenses as a brokerage fee, and the remaining 90 percent goes to the government. A moderate number of licenses have so far resulted from ITPA operations.

The private sector in Japan also has a strong voice in helping to shape the government's science and technology policy. One of the principal organizations representing business and industry is Keidanren. Established in 1946 through the merger of several economic associations, the Keidanren today has a membership of nearly 1000 firms and associations representing the nation's entire industrial, commercial, and financial sectors. Keidanren provides a forum for Japan's business community to consider common problems and to present business's views to the government on outstanding issues facing the nation and the economy. Keidanren promotes interaction across industrial sectors and also maintains close contact with other trade and economic organizations such as the Japan Federation of Employers' Association and the Japan Chamber of Commerce. Keidanren representatives are found on numerous government committees and advisory bodies as the government generally solicits its views in trying to reach a consensus on matters concerning policy planning and implementation. Keidanren favors liberalizing the nation's import restrictions on foreign capital and technology and raising the level of government investment in R&D. Although not all of Keidanren's recommendations are accepted by the government, the organization's views have been undeniably influential in helping to shape government policies and plans.

The aforementioned organizations fulfill important roles in Japan's structural make-up for the development and implementation of a science and technology policy that is fully supportive of national development goals. Familiarity with their mandates, functions, and interrelationships permits a better understanding of direct measures undertaken by the government to stimulate technological innovation, industrial development, and economic growth. These measures take the form of government subsidies and loans mostly for technologically based industrial-development projects. Government R&D funds are allocated primarily to universities and to government laboratories and institutes within AIST involved with big-science projects and the NRDP.[4,5] Low interest government loans also provide needed financing to spur the post-R&D commercialization process. However, Japanese industry is the principal performer and sponsor of R&D. For instance, in 1973 approximately two thirds of R&D expenditures came from the private sector as opposed to only 13 percent from government and 18 percent from educational institutions.[5,11] Hence the availability of government R&D funds is not a major stimulus for technological innovation in Japan. More likely the availability of government credits for industrialization projects has had a more profound influence on the nation's industrial progress as evidenced by the rapid growth of the JDB into one of the largest development banks in the world. The operations of the JRDC and the JITA also contribute in a direct way to spurring technological innovation. By providing investment capital for high-risk technological projects, the JRDC significantly aids in bridging the gap between research and commercial realization. In addition, the JRDC along with the JITA contribute importantly in their technology-linker roles. Through their conscious efforts to expedite the flow of technical information and technology, they create more favorable conditions for technological innovation and consequently considerably enhance the likelihood of successful commercialization.

The favorable climate for innovation in Japan also results in large part from a host of more indirect government measures that have an important bearing on private-sector investment decisions on whether or not to undertake R&D projects and industrial ventures. Prominent among these measures is the institution of a scientific and technological information policy designed to improve a firm's accessibility to technical information primarily during the early stages of the innovation process.[5] By this means crucial early-phase decisions on a particular innovation might be based on the best available information. An outgrowth of this policy was the formation of JAPATIC to serve as an information clearing house on foreign and domestic patents. Convenient access to patent information fosters the transfer and licensing of invention technology as well as helps to expedite the innovation process. The government's efforts in providing analyses of export markets and available foreign technologies also help to widen the store of information needed for industry to make better investment decisions, thereby significantly enhancing the likelihood of success in innovation projects. Another important measure is the promotion of industrial standardization by the government. Standardized products and equipments help to increase the utilization efficiency of existing industrial plant and equipment. Wider use of standardized equipment lessens the need for new capital investments, lowers development costs, and reduces technical risk in the industrial environment.

All these measures help to encourage innovation in Japan. However, perhaps the most pervasive and far-reaching measures enacted are the set of fiscal incentives designed to encourage R&D investment, modernization of industrial plants, and technological exports. The fiscal incentives benefit all industrial sectors and are intended to strengthen the firm's financial capacity for innovation.

A major incentive is the tax credit allowable since 1967 on a firm's R&D expenditures. If a firm's R&D expenditures in any given year exceeds the highest expenditure figure of any prior year since 1967, then 25 percent of the difference is allowable as a tax credit. If R&D expenditures exceed the previously high figure by more than 15 percent, then 50 percent of the amount in excess of 15 percent is granted as a tax credit. In any event the maximum credit that can be taken is 10 percent of the firm's total tax for that year. This tax measure is designed to encourage R&D investment and is particularly helpful in stimulating R&D activity in small companies.

Another measure enacted in 1958 provides for accelerated depreciation of plant and equipment during the first year. To take advantage of this tax provision, a special certificate of approval must first be obtained from MITI and the MOF. Applications for the certificate are evaluated according to factors such as whether the asset in question represents new technology, contributes to the national interest, promotes the nation's economic goals, or addresses a promising market area. After being granted the certificate, the firm is eligible for a first-year special depreciation of up to one third of the purchase cost of the new plant, equipment, or building (exclusive of land). This special depreciation is in addition to the normal depreciation entitlement of the firm. Firms benefiting from this depreciation measure are also able to claim a 50 percent tax deduction figured on the cost of the newly acquired assets during the first 3 years after acquisition. These tax provisions therefore constitute significant incentives for companies to industrialize and to modernize facilities.

Special fiscal attention is given to the small- and medium-sized companies because of their special technical and financial problems. The government has authorized additional tax benefits for small- and medium-sized firms over and above those available to large enterprises. These benefits include a 50 percent tax credit for R&D expenses and investments, a 20 percent first-year accelerated depreciation on newly acquired plant and equipment, a lower corporate tax rate compared to large companies (approximately 28 percent rather than the 36 percent typical of large firms), tax exemptions on reserve funds and local taxes, and exclusions from antitrust regulations under certain conditions.

Quasi-government corporations, R&D cooperatives, and research associations also come under special tax treatment. The quasi-government corporations operating under special charter are completely tax exempt. Companies in the mining and manufacturing sectors that form joint R&D ventures as cooperatives or research associations are permitted to depreciate fully their investments in fixed assets plus expenses incurred in setting up the joint venture. In addition, contributions that participating firms make to the new ventures are tax deductible. The intent of these measures is to encourage cooperative R&D and pooling of technical talents and resources among firms in the nonservice sector of the economy.

The above tax incentives serve to encourage private-sector investments in R&D and capital equipment that are important in the early, preproduction phases of the innovation process. Another tax measure designed to help draw innovations toward commercial success is a tax exemption granted on income from sales of technological products overseas. This particular measure serves the dual purpose of stimulating export trade and supporting the national goal of building a technology-intensive industry. This incentive is undoubtedly a strong factor in Japan having developed into the major industrialized, exporting nation it is today.

Comparative Assessment: West Germany and Japan

West Germany and Japan are clear testaments to the popular refrain that it is better to have fought on the wrong side during World War II, for both nations have risen from the ashes of defeat to take their places along with the United States as the leading economic powers in the Free World. Both nations have rebuilt their industries, rekindled the entrepreneurial spirit among their citizens, and achieved their spectacular economic growth within a very short time by historical standards. The means by which postwar development came about in the two countries are parallel in many respects and hold valuable lessons for other aspiring nations. A comparison of the areas of commonality and dissimilarity would help to identify those direct and indirect measures most effective in stimulating industrial innovation and economic growth. Before doing so, however, it would be useful to refer to Figure 1-5, which is a model of the technoeconomic interface. We will see that both West Germany and Japan employed measures aimed at stimulating all the different interface levels—science and technology base, technological innovation, and productivity growth.

In a general sense West Germany and Japan are very much alike in that they are not large countries and they both lack an abundance of natural resources. Both countries are highly dependent on export trade, and the average worker commands wages that are among the highest in the world. The two countries also benefit importantly from the industriousness and quality work ethic of their respective peoples. German workers are noted for their dedication to efficiency; likewise Japanese workers have a reputation for diligent hard work. Capital formation in both countries is aided by a comparatively high consumer savings rate. Their respective labor forces also generally get along well with management. Although the specific mode of labor-management relationship in each country differs, the end result is the same in that they provide a framework for cooperation and consensus rather than of contention between labor and management. A notable difference does exist, however, in that Japan, being a more populous nation, relies entirely on the labors of its citizens whereas Germany must import workers from neighboring countries to fill jobs shunned by its citizens. Since the mid 1970s, however, Germany's use of foreign workers to supplement its domestic work force has been slowly declining, apparently triggered by or at least coinciding with the recessionary world economic climate prevailing at the time.

By virtue of their unique geographical locations, postwar economic development in the two countries differed in one major respect. In Japan the postwar period was characterized by general anathema towards anything having to do with the military. As a consequence, the nation was able to devote its undivided attention and full resources towards reconstruction and building up the national economy. In contrast, after the war Germany found itself a country divided by the struggle between communism and democracy. Being literally on the frontlines of this contest, West Germany, as a major partner in the North Atlantic Alliance, necessarily had to maintain a sufficiently strong military capability to pose an effective defensive deterrence in Central Europe. It was therefore inevitable that a strong defense industry that is today highly respected even in the United States should develop in Germany. The West German defense industry has produced keenly competitive armament systems, particularly in aircraft, missiles, and tanks, some of which are being adopted in the American and NATO arsenals. The growth of a defense industry in West Germany has not in general detracted from the rapid development and growth of the national economy. On the contrary, close attention to potential technology transfers from the military to the civilian context helped to augment the development of the civilian economy. In fact the German government attempts to restrict its military R&D projects to those having high probability of civilian spin-off while projects having purely military applicability are generally performed elsewhere.[12] In this respect, Germany shares the widespread European view that successful transfer of military technology to the civilian sector must be consciously planned and pursued.

German and Japanese export trade reflect the differences in their industrial structure. Much of Japan's exports are decidedly consumer oriented while a sizable share of German exports include defense-related products. Both nations nevertheless are competitive in many areas. For example, the former German dominance in camera and lens production has been overtaken by the Japanese. Also, leadership in the American small car market has passed from German Volkswagens to Japanese Toyotas, Datsuns, and Hondas. Other export markets commonly served by the two countries are in chemicals and industrial goods such as machinery, and iron and steel products. Both Germany and Japan pursue trade policies that result in a high rate of technology transfers across their respective borders in both directions. Japan, however, has tended to impose closer government controls on its technological trade than Germany, although the recent trend has been towards a relaxation of governmental controls. Germany has pursued more of a free trade policy as evidenced by its higher volume of trade with the Soviet Bloc countries. It also imposes no restriction on the inflow of foreign capital investments into the country, whereas Japan has always regulated foreign investments in the country closely and only recently introduced a loosening of governmental restrictions.

Government relations with industry in the two countries are generally considered to be more closely cooperative than the arms-length, government-contractor relationship typically found in the United States. Government attentiveness to the needs of domestic industry in Germany and Japan is exemplified by the extraordinary efforts expended to assist industry in commercial ventures. For instance, government-industry cofunding of industrial projects is encouraged to assure in part the company's commitment to commercialize results upon the successful completion of the project. In addition, the governments of both countries compile data on world economic trends and trade and industrial developments, and they generally become involved in market research activities in direct support of domestic industries. West Germany has gone even further in trying to stimulate industrial innovation by making available public grants for introducing promising innovations into the commercial market. Also, through governmental risk sharing in venture-capital projects, the German government has attempted to lower risk-capital impediments to the innovation process. Both Germany and Japan also share a strong commitment for helping small- and medium-sized enterprises, as they recognize that future innovations and economic growth could well depend on paying close attention to ways of nurturing and strengthening the financial health of the newer and smaller companies. However, special attention for the small- and medium-sized firms does not extend to indiscriminate support and subsidization of weak firms. The principles of the free-enterprise system are not compromised simply to maintain employment. Companies that are not competitive simply go out of business; government attention and assistance are directed towards firms with more promising potential. This policy applies as much to large companies as to small ones.

It is also clear that both West Germany and Japan share a close identification of their national science and technology policy with economic growth. Recognizing that the science and technology base must be intimately coupled with industrial development to sustain future economic progress, both nations have instituted a system for R&D that has a strong industrial orientation. In both countries the major share of R&D is performed and financed by the private sector.[5,13] Of course, government-financed R&D projects are also an important part of the overall R&D picture in the two countries, but the industrial motivation behind these projects is generally unmistakable—whether it be in the context of direct commercialization of results, the generation of new industries, or simply developing commercial spin-offs of government R&D. To help assure that industrial and commercial benefits develop from their respective national investments in R&D, the two countries employ a wide participative system for R&D planning. Representatives from government, industry, and nonprofit institutions become involved in the planning process through numerous advisory groups and committees. The consensus reached then lays the foundation for mutual cooperation during later implementation stages.

Though the aims of German and Japanese science and technology policy are similar, the approach employed to implement the policy naturally differs in accord with the different cultures and institutional make-up of the two countries. In Germany, the government does not operate its own research laboratories but depends primarily on the different private, nonprofit research institutions and associations for R&D implementation. The government nevertheless is the main funding source and as such controls the direction of research activity for these nonprofit organizations. Also the German government attempts to gain maximum leverage on R&D expenditures by sponsoring projects in industrial research associations where results might be useful to a large number of companies. The nonprofit research institutions in a sense operate as intermediaries between the government and industry. They serve to aggregate the R&D needs of the private sector and to diffuse R&D results to the industrial enterprises. In contrast, besides research establishments in educational institutions and the private sector, Japan has a number of government R&D laboratories, such as those in the STA and the AIST. An important element of Japan's structure for industrial innovation is the banking sector, because Japanese industrial firms are generally highly leveraged by Western standards, and the banks are the primary source of credits for high-risk industrial ventures. The nation's banks therefore fulfill a role similar in many respects to the role of the private, nonprofit research institutions in Germany. That is, they are instruments of government policies for stimulating industrial development and economic growth. They differ in the sense that the banks' intervention in the innovation process generally takes place after the R&D stage and takes the form of low-interest loans rather than direct government subsidies.

Japan and Germany both share the common understanding that the innovation process requires more than simply an enlightened program for R&D. Considerable attention is consequently devoted facilitating innovation at all stages of the process. Special linker organizations exist in both countries specifically to lower the barriers to innovation. In Germany the Garching Instrument Company and the ARPAT both act to narrow the information gap between research and commercial applications. In Japan, the JRDC, JAPATIC, JITA, and the ITPA perform similar functions but with an expanded horizon that also includes foreign developments. These organizations, by expediting the flow of technical information, appreciably contribute to the stimulation of new ideas, the transfer of technology, and an increased likelihood of innovation success. The importance of capital formation in industrial innovation is also widely recognized although it is differently addressed in the two countries. The JDB in Japan plays a key role in providing development capital in Japan; on the other hand, Germany makes available government grants and partially underwrites risks in assignments of venture capital. In the patents area, commercial exploitation of patented technology is aggressively pursued in both Japan and Germany. However, Germany has instituted more additional measures to protect the rights of inventors and to provide greater incentives to encourage inventive activity. A broad array of fiscal incentives has also been established in both countries to encourage R&D activity and capital investment. Various forms of tax benefits are permitted individuals or corporate entities engaged in R&D or investing in industrial assets. These benefits range from tax credit allowed on R&D investments, to accelerated depreciation schedules on plant and equipment employed in R&D activity, to lowered tax rates for small- and medium-sized firms. Fiscal incentives have been found to be powerful tools for reducing the financial risks associated with all phases of the innovation process.

In comparing two of the world's leading industrial economies, the useful lessons to be learned must necessarily be deduced from the major points of commonality uncovered in the process. Reviewing the German and Japanese examples, several points of commonality can be found: (1) an industrious and frugal people, (2) a labor-management relationship founded on consensus and participative management, (3) a framework for close government-industry cooperation, (4) a free-enterprise commitment where weak, noncompetitive firms are allowed to go out of business, (5) a heavy industrial R&D emphasis self-financed by industry in the main, (6) an unimpeded, bidirectional flow of technology across national borders, (7) the existence of linker organizations to expedite the transfer of technical information and technology, (8) a broad system of fiscal and regulatory incentives to stimulate all stages of the innovation process, and (9) special attention to small- and medium-sized business enterprises. These various points of commonality individually and collectively serve to reinforce the technoeconomic interface be it through building a stronger science-technology base, accelerating the transfer of technology, or developing productivity gains, for example, through the introduction of newer and more efficient production processes. The obvious inference then is that these particular points of commonality constitute the principal factors underlying the economic success of West Germany and Japan during the postwar period.

CHAPTER **6**

The Climate for Innovation in the United States

The United States is world renowned for its inventiveness, innovativeness, and technological excellence. Major American innovations—such as xerography, instant photography, transistors, lasers, synthetic textile fibers, and the airplane—have had a revolutionary impact on today's modern society. Moreover, America's sparkling record of innovative achievements was built with minimal intervention from the government. The government's role was generally confined to specifying objectives (government procurements) and providing funds for R&D programs. A more forceful government role in the innovation process was unwarranted as long as American technological leadership remained unchallenged.

More recently, however, disturbing signs have appeared indicating a progressive erosion of America's traditional leadership position in technological innovation. A slower productivity growth rate combined with considerably greater competition from foreign manufacturers in domestic and international markets spotlight the need to revive America's former technological strengths and innovative ability. The productivity slowdown is symptomatic of a growing inability to utilize technology effectively rather than a lack of knowledge of it.[1] Increased foreign competition in commercial markets reflects in part the relative maturity of America's industrial machine compared to those of other nations, particularly West Germany and Japan, whose industries were largely rebuilt after World War II with modern facilities and the latest technologies. Associated with mature industries is a natural reluctance to change from established production methods to newer, more efficient processes when the changes involve large capital expenditures and severe disruptions in current production. Consequently, incremental im-

161

provements rather than revolutionary processes for productivity enhancement tend to be adopted in established industries. The combination of lagging productivity growth and imbalances in international trade competitiveness therefore contributes to a growing inflation that raises costs of R&D and production, reduces potential returns on investment, and generally acts to discourage would-be entrepreneurs from undertaking new, innovative ventures. All this points to the need for institutional changes in the United States if the nation's former innovative vitality is to be restored. What these changes ought to be has been the subject of much discussion and critical examination by both the public and private sectors of the economy.

Government-Industry Relationships

Ever since the industrial revolution the federal government has been concerned primarily with fostering conditions for economic growth and industrial development. Government attention was focused on protecting the integrity of the free-enterprise system. Economic forces and domestic market opportunities dominated the attention of industrial and business enterprises. Antitrust legislation was enacted to foster fair and open competition in the marketplace. The government's role in commerce was one of paternal detachment as free-market forces were allowed to flourish. Only in crises such as wars or threats to the national well-being did the government find it necessary to exert emergency authority to mobilize industrial resources in the national interest. In peacetime a close partnership between government and industry is maintained in the defense and space sectors, where the government constitutes the industrial market. But with the greater interdependence of nations and the escalating national socioeconomic priorities and issues, the government is forced to reexamine its traditional responsibilities and relationships with the private sector in light of increased international competition, the obvious need to strengthen the domestic economy, and the desire to improve the quality of life of the citizens. The issue is how working relationships between government and industry can be improved to meet the nation's domestic needs and its international commitments.[2]

In 1978 the Carter Administration consequently initiated a domestic policy review, eliciting the participation and views of leaders in industry, government, organized labor, the academic community, and the general public on how to strengthen industrial innovation in the United States. The policy review was directed by the Secretary of Commerce and supported by an

Advisory Committee composed of approximately 150 citizens who were divided into seven subcommittees. The focus was on how U.S. innovation could be improved to raise productivity and sharpen the nation's competitive position in world markets. Some of the principal policy issues considered were the need to (1) reduce regulatory uncertainty, (2) increase tax incentives, (3) institute reforms in the U.S. patent system, (4) introduce innovations education in institutions of higher learning, and (5) increase government-industry cooperation in developing generic technologies, that is, fundamental technological advances that herald downstream industrial innovations.[3] From this review came a broad government program to promote innovation in industry by means of improving technical information exchange to aid American industry in keeping abreast of foreign technological developments, encouraging the exclusive licensing of government-held patents, and pumping more R&D dollars into the small business community. Other steps to aid industrial innovation include using federal purchasing power to emphasize buying products of small and minority enterprises based on life-cycle costs rather than purchase price and improving business and technology education at universities and colleges. Considerations of new tax incentives to stimulate industrial innovation were deferred and remanded to further study in part because the introduction of significant tax breaks in 1979–1980 was regarded to be too inflationary in an already highly inflationary economy.

The developing consensus is that a national strategy is needed to bring into proper balance governmental incentives on the one hand and restraints on the other to produce a favorable climate for industrial innovation. The nation's growing social needs have necessitated a more active involvement by the federal government in regulating the business environment. The resulting plethora of federal and state regulations issued, some of which may be contradictory in nature, nevertheless tends to divert company resources from attractive investment areas and at the same time raises the cost of doing business. The projected social benefits of regulatory requirements must therefore be carefully weighed against the drawbacks that work to increase business risks and to reduce the availability of investment capital.

Existing laws and institutions also need be reexamined, taking into consideration changing patterns of international competition and the ability of American industry to meet the competitive challenge. In particular, the government's antitrust laws must be reviewed to determine whether domestic conditions prevailing at the time the laws were enacted are equally applicable in today's more interdependent world. This question is underscored by the increasing role of multinational companies in world commerce and the growing penetration of U.S. markets by foreign enterprises. Consider also the U.S. patent system, which was originally set up to provide the inventor exclusive rights to his invention for a limited time period in order to encourage its exploitation and commercialization. Patent protection was meant to provide the incentive to induce full public disclosure of the invention and its early introduction into commerce.[4] Inventions without protection tend to be kept secret and therefore serve to stifle commercial exploitation. Over the years, however, a number of difficulties have arisen, for example, uncertainty of the reliability of protection provided by a patent grant and the lengthy process associated with searching a patent application.[5] As a result, the U.S. patent system is viewed in many quarters as more of an impediment to innovation, and trade secrets are being increasingly used to circumvent the system in the case of keystone inventions. This of course defeats the original intent of the patent laws, which is to promote early public disclosure of inventions to expedite commercial exploitation.

Direct U.S. government interaction with industry is perhaps best typified by the arms-length, customer-contractor relationship. Government agencies contract for supplies and services from industry in support of their respective mission programs. Closer cooperative ties are to be avoided lest the cozy arrangements lead to conflicts of interest or to charges of collusion and favoritism. The customer-contractor relationship extends only over the life of the contract, after which the special relationship is terminated. While this formal relationship has worked well over the years in providing government with needed supplies and services and in promoting industrial progress, it has not fostered a deep understanding or mutual trust between government and industry. Government is suspicious of industrial motives and is wary of the political and economic power of the large companies. Industry on the other hand feels that government imposes too many restraints on business activities and is nonappreciative of the profit motive that underlies the free enterprise system. As a result, close cooperation to advance mutual interests is conspicuously missing. Seldom is industry invited to provide counsel during the inner workings of government planning. Industrial participation in government advisory councils and influence on government policies has waxed and waned with different administrations. By the same token government has hardly exerted strong leadership in areas of prime concern to industry—witness the absence of a public agency for promoting innovation in industry or for seeking to mitigate the market inroads made by foreign competition.

Although hints of a government-industrial complex is commonly viewed with distaste, there nevertheless is a general recognition that closer government-industry cooperation is desirable. However, the form and framework that such overall cooperation should take remains undecided. In the interim some progress can be noted in specific areas of mutual government and industry concern. One such area, generally considered to be an effective means to stimulate economic growth, is increased productivity in both the public and private sectors. The U.S. General Accounting Office (GAO) has in fact proposed that manufacturing productivity be established as a national priority. The importance attached to improving manufacturing productivity is reflected in the creation in 1975 of a public organization, the National Center for Productivity and Quality of Working Life, to serve as the national focus and to take the lead in formulating a national policy. The priority assigned to productivity improvement remains a high one despite the difficulties experienced by the National Center on Productivity that eventually led to its abolition in 1978. However, responsibility for carrying out specific programs for enhancing productivity is still assigned to the different mission agencies. As it turns out, the Department of Defense has the largest program. Roughly $165 million is earmarked annually for industrial projects sponsored by the three military services designed to develop advances in manufacturing technology. The principal objective is to achieve more efficient production and thereby lower defense procurement costs.

Though substantive progress can be identified in specific programmatic areas of both governmental and industrial concern, there is no unifying national strategy or policy to provide the bonding agent for the disparate programs. Consequently, the modest degree of government-industry cooperation that does exist is disjointed and limited in scope. This state of affairs is in large part due to the decentralized character of the federal government and the relative autonomy of the different mission agencies. Each agency retains the management prerogative to formulate requirements and establish programs designed to carry out its mandated function. Though coordination and cooperation with parallel agencies in areas of overlapping responsibility is developed in many cases, it must still be recognized that no mission agency has the resident authority to formulate a national policy that also affects other mission agencies. Because deliberations on national policy to stimulate government-industrial cooperation and industrial innovation affects virtually all the mission agencies, the burden of leadership falls on none but the White House. Unfortunately, high-level deliberations on these matters have a habit of being preempted by other more pressing issues.

It is no small wonder then that industry has turned to influencing the governmental process mainly through the legislative branch. Industry as well as other special interest groups have found Congress to be fertile territory for lobbying efforts intended to gain special advantages or to ameliorate pending adverse legislation. Power brokers and professional lobbyists concentrate on influencing the key members of Congressional committees and subcommittees where debates on proposed legislation have special significance to their clients. Additionally, different sectors of business and industry are organized under various trade associations that represent the collective interests of their respective memberships on Capitol Hill and elsewhere. Some of the larger corporations also retain their own representatives and lobbyists who look after the individual company's special interests. Industry thus employs an army of surrogates on Capitol Hill that works to make industrial views known and to influence legislation through the art of skillful politicking and gentle persuasion.

Although industry's relations with the government have been rather diffuse, this is not the case with respect to labor organizations. Modern-day labor-management relationships were shaped in the 1930s with the passage of laws such as the National Labor Relations Act of 1935, which permitted workers to form unions and required management to bargain with the unions. The labor unions over the years have grown and acquired considerable power in their ability to call crippling strikes and boycotts that not only affect individual companies but also entire industries and even the national economy. Collective bargaining between managements and the unions are generally over clear-cut issues concerning wages, benefits, and working conditions. Companies entertain no doubts as to who the labor principals are at contract-negotiation time. More often than not the positions and arguments of labor and management are clearly drawn, and negotiations proceed in an adversarial manner. Particularly acrimonious negotiations unfortunately have a tendency to extend past contract settlement and could adversely affect worker performance and motivation. Seldom does the labor-management relationship include cooperative programs that extend beyond the collective bargaining process to the point where labor and management might share the benefits from joint efforts to increase productivity and profits.

Science and Technology Policy Considerations

The industrial environment is influenced by many external forces not the least of which is the nation's science and technology policies, which affect to a large extent the innovativeness and competitiveness of industry. Consequently, the industrial system must be ready to accommodate the ebb and flow of these external influences if it hopes to remain competitive. Although the need for structural changes in America's innovational system is widely recognized, the actual prescriptions for change are not easily determined as no precedence exists for active government intervention to guide conscious attempts to stimulate industrial innovation in the United States. The role that government should play and the degree of intervention desired are substantive questions that need to be resolved. The situation is compounded by the fact that U.S. policies (or nonpolicy) relating to science and technology and industrial innovation generally evolved over several government administrations since the conclusion of World War II.

During the war the nation's science, technology, and industrial might was harnessed in support of the war effort. After World War II, many of the nation's technological and industrial resources were redirected towards the creation of new technologically oriented industries responsible for major innovations in the aerospace, electronics, and nuclear and chemical fields.[6] During the Eisenhower Administration, federal investments in R&D grew rapidly primarily in the defense, atomic, and space sectors. The government specified R&D objectives, supported the projects, and was the principal user or customer. No attention was given to potential civilian spin-offs until the Kennedy Administration when in the early 1960s the question of the government's proper role in helping to reinvigorate American industry came to the fore. Discussions on the subject subsequently led to the emergence of technology transfer to the civilian sector as a major public policy issue by the mid 1960s. This was a time when people were concerned with generating greater social benefits from the nation's investments in science and technology. Then in 1972 President Nixon in the first Science and Technology Message to Congress called for the increased utilization of federal R&D results for meeting civilian needs.[7] Attention to civilian priorities was reinforced in the 1970s by the public concern for environmental protection, pollution control, more stringent health and safety standards, consumer product safety, more efficient urban transit systems, and other issues affecting the quality of life. The growth of government regulations and standards for these socioeconomic priorities had the effect of further complicating policy issues relating to technological innovation.

The complexity of issues affecting technology, innovation, and socioeconomic priorities in the United States makes the need for a clear and coherent science and technology policy all the more important. However, the science and technology picture in America is more a conglomeration of many disjointed policies that address specific issues and have evolved over a period of time than a coherent, well conceived expression of goals and objectives for harnessing science and technology in the national interest. In general, the government view has been that the development and exploitation of industrial and commercially relevant technologies is best left to the private sector. Furthermore, the self-interests of the individual firm in the marketplace will help assure the effective diffusion of proprietary technology to other firms for wider commercial exploitation.[5] Except in times of crisis such as World War II or the threat of Soviet superiority in space technology, the government has found it unnecessary to intervene more forcefully in the private sector. The intent, of course, is to preserve free enterprise in its truest sense.

The pluralistic development or, more precisely, evolution of U.S. science and technology has resulted in policies that are fragmented and in some cases even contradictory.[5] Consider, for example, U.S. export policies. The main strength of the U.S. export trade is in technology-intensive products. American economic interests therefore dictate strong promotion of technology-intensive exports. However, American security interests dictate that more effective controls be imposed on U.S. technology exports. Another area of contradiction concerns U.S. antitrust laws. In protecting American firms from predatory monopolistic practices, the laws also effectively hamper American competition against foreign monopolies and state-controlled enterprises in the international market. The antitrust laws also discourage the formation of cooperative R&D ventures, particularly among the larger high-technology companies where such joint ventures would be desirable to spread risks in expensive, high-risk R&D projects. As a final note, consider the current proliferation of government regulations that are being introduced at a time when the government is seeking ways to promote technological innovation. These regulatory barriers in fact act as strong disincentives for entrepreneurship and technological innovation.

The absence of a national policy on science and technology is perhaps not so surprising given the decentralized character of the federal government. The nation's founding fathers were careful to build into the Constitution a system of checks and balances where power is divided among the executive, legislative, and judicial branches of government. Within the executive branch are many agencies and departments established to fulfill specific mission responsibilities such as national defense, space exploration, transportation, and energy. Each of these agencies and departments initiate their own R&D programs to carry out their respective mandates. More often than not these programs are poorly coordinated across interagency lines. Similar

R&D work may be undertaken and test facilities built where the sole distinction may be only in the eventual mission application or constituency served. Though virtually all the federal agencies conduct research in one form or another, the largest R&D efforts are in the defense, space, energy, and health areas. The Defense Department and NASA are beneficiaries of their own R&D programs since they are also the ultimate customer or user for defense and space products, respectively. In fulfilling their primary mission responsibilities, the agencies are not obliged to coordinate programs with other federal agencies. However, some agencies, such as NASA, the Environmental Protection Agency (EPA), the Departments of Transportation (DOT), Agriculture (DOA), and Energy (DOE), are explicitly chartered to transfer technology. However, the transfer is typically to their respective public constituents rather than between agencies. The agency best positioned to coordinate across interagency lines is the Office of Management and Budget (OMB), but its principal role is to be the conduit for congressional appropriations for the different executive agencies. OMB does not coordinate agency programs across the executive branch.

The importance of technological innovation to the economic health of the nation is widely recognized within the federal government. However, because of the clear-cut distinctions in the mission responsibilities of the executive branch agencies, no agency has assumed the responsibility to promote technological innovation in industry, nor has any new government organization dedicated to fostering industrial innovation been established.

The resultant leadership vacuum has not helped allay private-sector uncertainties on matters relating to science and technology policy. These uncertainties in turn adversely affect the extent of private-sector investments in R&D and technological innovation projects. Institutional changes have been slow in coming not because of a lack of will but because of insufficient knowledge as to the nature of the corrective action required. After all, conscious, concerted attempts to reinvigorate the nation's innovative fiber basically constitute a pioneering effort never before undertaken under similar circumstances. As a result, the United States finds itself first having to identify workable policy options for stimulating technological innovation in the current economic and political environment. The domestic policy review ordered by President Carter in 1978 is one attempt to identify these policy options with a giant assist provided by the business community.

The basic question concerning the extent of government intervention needed and desired to strengthen innovation in America is practically speaking a moot point in relation to defense, space, and (to a limited extent) nuclear energy because the respective mission agencies—DOD, NASA, and DOE—represent closed-loop procurement systems. That is, these agencies sponsor R&D and are also buyers of the final product.* Hence the role of the federal government is pervasive in guiding and supporting all phases of work from basic research through product development, test, and evaluation to procurement and operations. It should also be noted that these agencies are three of the most heavily technologically oriented agencies in the federal government. However, the question regarding the proper role of government in spurring innovation emerged mainly as a result of the attention first spotlighted in the 1960s on how to strengthen civilian-oriented R&D for achieving socioeconomic gains. The question therefore is really associated more with commercially relevant technologies than with defense, space, or nuclear weapons technologies. This is not to say, however, that DOD, NASA, and DOE should be removed from government efforts to stimulate innovation in the civilian economy. On the contrary, these agencies have a definite role to play by virtue of their sizeable science and technology base and the fact that an appreciable share of defense, space, and nuclear weapons technology is transferable to civilian applications. All three agencies are in fact already engaged in efforts to transfer technology to the civilian sector, with DOD being the only one of the three not specifically mandated for this function.

In commercially relevant technologies the federal government has generally not been required to augment market forces for stimulating innovations in the private sector. With few exceptions, notably in agriculture and energy, the role of the government has been limited to supporting basic research and exploratory development of emerging technologies. It has left the process of technology exploitation and commercialization to the private sector. Private enterprises, as a consequence, have focused their efforts on applying technological know-how to meet market demands. Except for some major corporations and philanthropic organizations, industry seldom invests in R&D without a clear market objective and an adequate projected return on investment. Private-sector investment capital is mainly earmarked for the downstream phases of the innovation process, where costs are considerably higher than those encountered in the upstream R&D activity. However, in big science projects, particularly in energy R&D (e.g., ocean thermal energy conversion, fast-breeder reactors and synthetic fuel processing from coal), the government has found it necessary to continue active support and in-

---

*This is only partially true in the case of DOE. It applies to that part of DOE concerned with the development and production of nuclear weapons.

volvement past the basic R&D stage until commercial feasibility has been demonstrated at least through the pilot plant stage. This deeper involvement of the government in commercial-technology projects has raised additional policy issues concerning the transition of these projects from major federal involvement to full commercial implementation by industry. Resolution of these issues must necessarily consider the potential socioeconomic benefits to the nation from greater federal involvement in commercial technologies, the ability and inclination of the private sector to invest its own resources, and the range of federal incentives or disincentives already bearing on the matter.

Clearly, the minimum level of government involvement needed to strengthen the nation's innovative fiber is to mount conscious and concerted efforts to make the results of federally sponsored R&D more easily accessible to the private sector. Since 1953, more than $350 billion has been spent on federal R&D, but only an insignificant portion of the R&D investment is spent to transfer the resulting technology to the commercial sector.[8] For instance, in 1973 only about 1.5 percent of federal R&D expenditures was used to promote technology transfer and utilization. Of this share, the Department of Agriculture accounted for most of it through their agricultural extension service.† More typically, however, the government-owned technology reached the commercial sector primarily as a result of natural diffusion and serendipity. Invariably the burden fell on the private sector to uncover promising government technologies and exploit them in commercial markets. Recognizing that commercial applications may be developed for a technology in markets completely unanticipated at the time of its development, more active government efforts to promote the transfer of publicly held technology to the private sector should therefore help stimulate imaginative, intersectoral thinking and ultimately lead to a higher technological innovation rate for the nation.

A more active federal role in stimulating the transfer and utilization of government technology for achieving socioeconomic benefits has in fact been the subject of considerable attention. A committee of the U.S. National Academy of Engineering (NAE) looking into the subject in 1974 concluded that technologies rarely find spin-off applications on their own but generally require conscious efforts to transform and adapt them to new uses. Radar,

---

†The DOA extension represents perhaps America's earliest and most successful attempt to transfer actively government-developed research results to the civilian user. The service provides assistance to the nation's farmers in adopting the latest in agricultural advances and know-how by employing a nationwide network of trained field agents.

antibiotics, and the jet aircraft are some of the well-known examples of innovations that originally came from mission-oriented government projects that later attained widespread commercial acceptance without too much difficulty. But the NAE contends that these examples are more the exceptions than the rule. In surveying 25 federal agencies, the NAE found a serious imbalance in the attention and funds allocated for technology transfer and utilization compared with their overall R&D budgets. Additional problems cited were the absence of a proper legal mandate for technology transfer in most of the agencies surveyed as well as the lack of personnel billets and proper government job classifications for staff engaged in technology transfer activities. Accordingly, the NAE recommended that more federal funds be budgeted for technology transfer activities and that the mission agencies be provided the legal mandate and staff resources for these activities.

In a separate investigation in 1972, the GAO, responding to presidential and congressional concern that the nation's R&D investments are not adequately addressing critical social problems facing the nation, looked into the question of using defense technology to help solve pressing national problems in the civilian sector.[9] The DOD was singled out for the GAO investigation because annual defense R&D expenditures amount to nearly half the total federal R&D budget. Furthermore, defense R&D spans virtually every scientific and technical discipline. The GAO concluded that DOD should more actively pursue the transfer of technology to the civil sector. It recognized that the profit motive was sufficient in the past to cause the private sector to commercialize defense-based technologies such as computers, communication satellites, inertial navigation systems, the use of lasers for eye surgery, and infrared sensors for fire detection. However, more attention was deemed necessary to solve urgent social problems such as in transportation, law enforcement, and urban development because the potential markets in these areas were not sufficiently aggregated to attract private enterprise. The GAO identified the key constraints to a more active DOD technology transfer effort to be (1) the lack of a DOD policy for technology transfer, (2) the mistaken interpretation by some DOD officials that existing legislation originally intended to constrain the use of DOD research funds to mission-related projects also applies to the transfer of research results to nondefense applications, and (3) the justifiable concern that staff resources devoted to nondefense projects will lead to reductions in authorized personnel ceilings. The GAO recommended that DOD establish policy guidance for technology transfer and that OMB do likewise for the government as a whole in order to stimulate more active government efforts to transfer technology to the civil sector.

In 1974 the Federal Council for Science and Technology (FCST) completed a study to determine how federal R&D centers and laboratories could be better utilized in providing technical assistance to governmental organizations beyond their parent agencies, specifically, state and local governments and other federal agencies.[10] The federal R&D field activities constitute a considerable national resource comprising some 469 installations and 11,900 professionals spread out in all ten federal regions of the United States. This far-flung system of laboratories falls administratively under the different federal mission agencies. As a result, the federal R&D laboratories primarily serve the mission of their parent agencies with little attention given to solving intergovernmental problems. Despite the absence of legal or policy constraints, the existence of an array of institutional barriers acts to discourage more effective intergovernmental use of the R&D field activities. Principal among the barriers are (1) the balkanized structure of the federal laboratory system, (2) tight manpower and budget ceilings, (3) strong mission priorities that supersede nonmission-oriented projects, (4) a diverse, nonaggregrated user community represented by 50 state governments and thousands of local jurisdictions and communities, (5) the existence of information gaps and communication difficulties between R&D organizations and state and local governments, and (6) the lack of incentives for laboratory managers to engage in nonmission-oriented work. To improve intergovernmental cooperation, the FCST drafted a policy statement to expand interagency use of federal R&D centers and laboratories. Among its recommendations was the generation of increased use of the Intergovernmental Personnel Act of 1970, which authorizes temporary exchanges of professional personnel between the federal, state, and local governments. Personnel exchanges would help break down communication barriers and would provide an effective mechanism for technology transfer via the skills and know-how of the individuals involved.

Throughout the 1970s considerable attention was also given by Congress to issues relating to technology transfer and innovation. Several bills have been introduced in both houses of Congress, and committee hearings have been held on various topics and issues mirroring the national concern on how best to marshall the nation's vast scientific and technological resources towards solving current national problems in the socioeconomic sphere. Also in response to the national concern, President Carter initiated the domestic policy review in 1978. Associated with the review came the concept of cooperative technology centers proposed by the Department of Commerce and patterned after cooperative centers such as those found in Japan. These centers are envisioned to be centers of excellence in different areas of industrial technology such as powder metallurgy, industrial coatings, composite materials, and semiconductor fabrication. To be jointly funded by government and industry, cooperative technology centers would give balanced emphasis to industrial R&D and technology transfer, and in so doing are expected to compress significantly the average U.S. innovation time period.[11]

The small business community has been given particularly close policy scrutiny because small business is recognized to be a principal source of major innovations when compared with big business, government laboratories, and universities. Small businesses have made disproportionately large contributions to productivity increases, job creation, higher tax revenues, and product innovations. For example, an M.I.T. study based on a 5.6 million person sample found that, between 1969–1976, 66 percent of all new jobs created in the private sector came from firms of 20 or fewer employees, while 80 percent of the new jobs came from firms with 500 employees or less.[12] However, the contributions and innovativeness of the small firms have declined in recent years, primarily because of inhibitive government policies in the view of many. As a result, several independent task forces were formed aside from the President's domestic policy review to study the problem and to make policy recommendations.

The Department of Commerce through the Commerce Technical Advisory Board (CTAB) in 1978 commissioned a study group to develop policy recommendations for the creation of more jobs in small, innovative technology-oriented firms. The group identified the root causes of difficulties faced by small businesses as (1) higher capital gains taxes that greatly reduced capital availability, (2) increased regulatory requirements that made it more difficult for small firms to raise capital, and (3) concentration of government R&D generally in the larger firms and typically in a few industries.[13] Solutions proposed by the CTAB group were directed towards (1) increasing capital availability by means of a reduction in the capital gains tax rate and a more advantageous corporate tax rate for small businesses, (2) cutting costs of compliance with government regulations by allowing higher tax deductions, (3) stimulating the small business exploitation of technologies developed in government laboratories and large companies through appropriate changes in the tax laws and greater federal investments in technology transfer, (4) stimulating R&D in small businesses by means of higher allocation of federal R&D funds for the small business sector, and (5) strengthening export trade through creation of new international trading organizations representing small businesses.

At about the same time that the CTAB study was commenced, the Small Business Administration (SBA), Office of Advocacy, commissioned a task force of small business executives and venture-capital managers to look into how best to strengthen small, innovative technology-oriented firms. In a parallel development, the small business representatives that were part of the subcommittee activity of the domestic policy review came together and prepared a supplementary small business report to document more clearly the small business perspective and problems. This was deemed desirable because the large majority of business representatives participating in the domestic policy review were from large corporations. The conclusions of all three groups—the SBA task force, the CTAB study group, and the small business participants in the domestic policy review—were in general agreement. In addition the task force favored changes in the Securities Exchange Commission's registration requirements for new stock issues. A new, simplified filing form was proposed for new stock issues under $5 million designed to facilitate the public sale of stock offerings from small businesses. Liberalization of the patent laws was also advocated so that small businesses could be given an opportunity to retain patent rights on inventions made under government sponsorship. Based on the findings of all three groups, the SBA task force promptly proposed new legislation entitled, "Small Business Innovation Act of 1979," portions of which have been introduced in the Senate Judiciary Committee and the Senate Small Business Committee.[14] The legislative proposal therefore combines the recommendations and advice from all three citizen groups on how best to enhance the environment for technological innovation in the small business sector.

The developing consensus is that greater government involvement is needed to expedite technological innovation in the United States. To what extent and in what form this involvement should take has yet to be determined. One thing is clear, however. Greater federal investment is being made in civilian R&D, that is, in fields outside defense and space. During the 1969–1976 period, federal spending for civilian R&D climbed at an annual rate of 16 percent, while defense R&D grew by 3.4 percent per year and space R&D declined by 2.9 percent annually.[15] However, there is a growing conviction that more R&D funds are not necessarily effective in spurring innovation. In attempting to determine viable policy alternatives, the federal government has devoted considerable effort to acquire a better understanding of the innovation process in general and the key implications for the United States in particular.

Leading this effort is the National Science Foundation (NSF), which for many years has sponsored research on the nature of the technological innovation process and the important environmental factors affecting technological change. Innovation research was initiated in 1972 with the institution of two programs within NSF, the Experimental R&D Incentives Program and the National R&D Assessment Program, designed to find ways to stimulate innovation and to determine appropriate federal policies towards this end. The objective of the Experimental R&D Incentives Program was to identify and test different incentives that the federal government might employ to promote technological innovation. The methodology employed was to identify the barriers to innovation, identify incentives for overcoming the barriers, and to design experiments to test the validity and effectiveness of the incentives. At the same time that the Incentives Program was started at NSF, a separate but similar program, the Experimental Technology Incentives Program (ETIP), was also established by Presidential mandate within the National Bureau of Standards (NBS) of the Department of Commerce. The objective of this program was to find ways to stimulate civilian R&D and the application of R&D results. ETIP mapped out five specific areas of emphasis: (1) civilian R&D policies and policy alternatives, (2) instituting incentives to innovation through the federal procurement process, (3) government regulations and their impact on innovation, (4) capital availability for small technology-based firms, and (5) federal financial assistance to industry and state and local governments to foster technological change.[16] In a general sense both the NSF Incentives Program and the NBS ETIP were designed to identify and test federal policy options for stimulating technological innovation. The NSF National R&D Assessment Program was also established for this principal objective but focused more on research and analyses of (1) the relationship between R&D and technological innovation, (2) the important factors underlying this relationship, and (3) the implications of alternative federal policy options for future R&D and technological innovation.[17]

In 1976 the National R&D Assessment Program was combined with other NSF activities concerned with U.S. science and technology policy into a new division for policy research and analysis. This change effectively concentrates science and technology policy research activities within a single office in support of the President and Congress. Research activities are divided into four principal topical areas: (1) socioeconomic effects of science and technology, (2) innovation processes and their management, (3) the role of science and technology in meeting national environmental, energy and resources goals, and (4) the direct and indirect economic, social and institutional consequences of technological change.[18] Research results are intended to provide federal policymakers with an improved knowledge base for making decisions relating to R&D and technological innovation. A number of studies have been conducted internally and by means of extramural awards, but definitive results as well as their realization in the form of new federal policies for stimulating innovation are generally not anticipated in the near future.

Government Efforts to Spur Innovation

Concurrent with the federal government's efforts to establish a policy framework for science, technology, and innovation that is responsive to present-day socioeconomic needs, a number of government programs and initiatives have in fact been introduced within existing policies and institutional arrangements designed to develop fuller exploitation of the nation's science and technology base for the common good. These pragmatic measures in the main are concerned with enhancing the transfer of technology as a means of spurring technological innovation, which is perhaps all that can be reasonably attempted in the absence of a more definitive understanding and agreement on needed public policy changes and institutional realignments. Not surprisingly, given the decentralized nature of the federal government, the different technology-utilization programs were with few exceptions established and organized under the different mission-agencies of the federal government. Of course, their success depends in large part on the willingness and ability of potential users in both the public and private sectors to adopt and apply the fruits of already-developed technology. Nevertheless, the programs in the aggregate represent a growing commitment on the part of the federal government to develop fuller utilization of the nation's science and technology resources. A brief look at some of these agency efforts should further help to characterize the climate for innovation in the United States.

Probably the earliest and most successful example of technology development and transfer to commercial use has been in American agriculture. The bedrock of this system is the Department of Agriculture's Cooperative Extension Service, which was established in 1914 to make available to the American people the results of agricultural research developed at universities, government laboratories, and other facilities.[19] The Cooperative Extension Service is a part of the federal land-grant university system and is administered by a land-grant university in each of the nation's 50 states. This agricultural extension service accounts for the major share by far of federal expenditures for technology transfer and utilization. The most crucial element in the service is the nationwide network of field representatives who provide an effective linkage between research results and the problems of the individual farmer. The information flow is two way, on the one hand providing farmers with the latest in agricultural technology and on the other hand relaying the farmers' technological problems to the researchers.[20] There are more than 3000 offices nationwide, and approximately 17,000 people working full time on extension activities are supplemented by more than a million volunteers. Extension agents and specialists constitute the core of the field network. The extension agents live and work among the farmers, agricultural industries, and other community groups and are intimately familiar with their agricultural problems. They serve to link the user to the various information sources in government and colleges. Backing up the agents are some 4200 specialists at the regional, state, and national level who supply the key interpretative link between researchers and the agricultural user. The unqualified success of the agricultural extension service is attested to by America's impressive record of agricultural trade surpluses registered year after year. Leadership in agricultural exports on world markets is maintained principally because American farmers are the most efficient in the world as farm productivity in recent years has risen considerably faster than industrial productivity. These achievements are attributed in no small way to the nation's agricultural technology delivery system built up over the years that links the latest technological advances with more productive farm output.

Space technology has also been the object of active efforts to develop greater civilian uses in both the public and private sectors. The National Aeronautics and Space Act of 1958, which established NASA, included a provision for a technology transfer function that led to the creation of the Technology Utilization Program in 1962. The principal objective of the program is to develop greater use of space technology in civilian government areas and in the commercial market in order to increase the returns from the nation's investment in space research and exploration. The main thrust is to promote the rapid diffusion of technical information on new aerospace technologies to both government and industry. To a limited extent NASA will also assist in the adaptation of specific technologies to civilian uses. The approach employed consists of (1) publication and dissemination of one-page summaries of inventions, technological developments and innovations considered promising for civilian applications, (2) establishment of six industrial application centers that provide information retrieval services and technical assistance in matching industrial problems with NASA technology, (3) creation of biomedical and technology application teams at research institutes and universities to define and seek solutions for important problems in the biomedical field and in the public sector, (4) institution of the Computer Software Management and Information Center (COSMIC) to make available to industry at nominal cost computer programs developed by NASA and other government agencies, and (5) providing adaptive engineering services that generally involve the redesign and reengineering of individual space technologies to tailor them to the specific civilian application. The approach therefore encompasses technology publication, identifying user requirements, and the necessary brokerage and technical assistance activities to bridge the source-user gap. Additionally, NASA enters into joint projects with other federal agencies where aerospace technology would play a role in meeting common objectives. To further spur innovations in industry based on aerospace inventions, NASA patent policy has been changed to relax the conditions by which exclusive licenses can be granted. In appropriate cases exclusive licenses may now be granted as early as nine months after the filing of the patent application. An important aspect of the NASA Technology Utilization Program is that it is geographically decentralized. NASA R&D laboratories, the industrial application centers, and the technology/biomedical application teams are located nationwide, thus facilitating considerably the identification of state and local government problems as well as the industrial needs in different parts of the country.

In response to growing national concern for developing tangible socioeconomic benefits from the nation's investment in R&D, the Department of Defense and the military services in the early 1970s became more actively involved in military-civilian technology transfer. Although this function was not legislatively mandated, the DOD nevertheless issued departmental policy memoranda in 1972, 1974, and 1978 encouraging the military services to engage in technology transfer activities to the civil sector. The objective is to develop greater returns from the defense R&D dollar by more actively seeking the application of existing military technology to civilian needs in both government and industry. However, these activities are not to impede the primary mission of the DOD and its components, and the level of effort in the DOD R&D centers and laboratories is not to exceed 3 percent of the professional staff. Furthermore, costs incurred are to be reimbursed by the beneficiaries of the particular work except when common military-civilian requirements dictate that cooperative development projects be established. In this case, joint funding is permitted with a civilian agency. Implementation of DOD policies for military-civilian transfer is delegated to the military services. Inasmuch as a considerable proportion of defense technology resides in the Services' R&D centers and laboratories, these field activities became actively engaged in seeking civilian applications for their resident technologies. A consortium of Army, Air Force, and (mostly) Navy laboratories was formed in 1971 to facilitate interlaboratory information exchange and cooperation in addressing civilian problems and issues. This DOD Laboratory Consortium eventually expanded from its charter membership of 11 field activities to become the Federal Laboratory Consortium for Technology Transfer with a membership of more than 180 federal R&D centers and laboratories embracing 10 federal agencies.[21]

Despite the fact that many important innovations in use today were originally derived from the results of military R&D, implementation of DOD's policy on military-civilian technology transfer created increased awareness and stimulated conscious efforts within the military services to develop civilian applications from military technology. Not surprisingly, the approaches taken and the level of activity differed among the three services. The most active efforts to transfer technology to the civil sector is found in the Navy program. The principal objective of the Navy Technology Transfer Program is to expedite the use of Navy technology for improving productivity in the public and private sectors. More recently, increasing attention has also been given to transferring technology from non-Navy sources to complement Navy inhouse R&D resources. The Navy played the leading role in forming the DOD Laboratory Consortium for Technology Transfer in 1971. Person-to-person interaction, a key ingredient in successful technology transfer, is fostered by the establishment of technology transfer focal points at about 65 different Navy R&D field activities across the nation. Navy personnel have also been placed on temporary assignments with civilian federal agencies, state governments, and local communities to facilitate the problem definition and communication process that is so important prior to the technology search and transfer phases. A major vehicle for bringing promising Navy technology to the attention of industry is the Navy Technology Transfer Fact Sheet, a monthly publication giving concise summaries of Navy innovations and technological advances considered to have high commercial potential. Industrial enterprises are encouraged to seek licenses or otherwise acquire the technologies for commercial purposes. Technical information for industry is also available through the Navy Research and Development Information Centers (NARDIC). These centers organized along with the Army and Air Force provide industry with information on R&D plans and requirements of the military services. More recently, the Navy has been placing greater emphasis on transferring technology to the small-business sector and in particular the small, technology-oriented enterprises.

Closely associated with the military-civilian technology transfer effort is the DOD Manufacturing Technology Program. The program, given strong impetus in 1975, is designed to develop improvements or new advances in manufacturing techniques, processes, materials, and equipment in order to develop timely, reliable, and economical production of defense systems and materiel. By improving the manufacturing productivity of defense contractors, substantial reduction in defense systems acquisition costs can be achieved. Special emphasis is placed on improving manufacturing productivity because manufacturing represents the largest single segment of the American GNP(approximately 30 percent). The total annual costs for machining in the manufacturing sector alone are estimated to be $60 billion.[22] Consequently, even a small percentage cost reduction would amount to a sizeable dollar savings. Overseeing the total program resides in the Office of the Secretary of Defense, but each of the three military services are given central management and control authority for their respective programs, which typically involve hundreds of separate projects contracted to industry. Because the potential benefits from the DOD Manufacturing Technology Program could be substantial for industry and the American economy in general, particular attention is given to diffusing the results of the program throughout the civilian sector. One important method employed is to require the industrial contractor on all projects to perform an end-of-project, full-scale demonstration in the presence of other industrial concerns including direct competitors. These demonstrations help disseminate the results of the projects. They are particularly valuable to other firms because the projects generally address generic manufacturing problems resolution of which would significantly lower overall manufacturing costs. Other means are also employed to speed the diffusion of manufacturing advances throughout the economy such as through industrial seminars and publication in journals and fact sheets. The importance of manufacturing technology transfer is underscored by the fact that modern manufacturing techniques tend to be concentrated only in the large, high-technology companies while small manufacturers, making up more than 95 percent of the manufacturing capacity in the United States, rarely have the opportunity to share in the higher production efficiencies derived from modern manufacturing technologies. A more complete assimilation of advanced manufacturing techniques throughout the manufacturing sector would have a large impact in boosting overall productivity in the national economy.

Certainly, federal agency efforts in technology transfer to stimulate innovation are not confined only to the technology-oriented agencies. The Department of Commerce through its diverse organizational components is engaged in a wide array of activities designed to promote technological innovation and economic progress. The agency's major technical arm is the NBS, which through its standards expertise and extensive measurements capability influences to a large extent industrial practices relating to production, quality control, parts interchangeability, and the use of standard reference materials. NBS R&D resources are also made available to other federal agencies through which it becomes involved in civilian R&D projects of national concern such as in energy, materials conservation, environmental pollution, and radiation safety. Another arm of DOC with no R&D mandate but nevertheless constituting an important factor in fostering technology transfer and innovation is the National Technical Information Service. This organization serves as the federal government's central repository for unclassified technical reports generated by the different federal agencies. Chartered to be self-sustaining, NTIS sells technical documents and other information products to all interested customers including foreign citizens and governments. NTIS also actively promotes the licensing of government patents domestically as well as overseas through publications, invention seminars, licensing exhibits, and direct personal contacts. Before filing government patents in foreign countries, foreign rights to the inventions are first transferred to NTIS from the originating federal agency. Royalties accruing from foreign licenses are returned to the U.S. Treasury. Since 1977, however, an inventor's incentive awards system was introduced permitting the government inventor to receive a small share of the annual royalties. The technical information services provided by NTIS have been recognized to be but passive mechanisms to promote technology transfer. Efforts to develop a more active technology transfer effort were made by NTIS in 1976 in cooperation with the Economic Development Administration (EDA) to complement existing technical information services and invention licensing activities.

The EDA was established within the DOC in 1965 to stimulate the development of economically depressed regions of the United States. The aim is to help states and municipalities create job opportunities and new industries. Management, financial, technical, and marketing assistance is also provided for private enterprises to help create jobs, increase local income levels, and raise investment capital for plant modernization and expansion. The primary vehicle is the EDA University Centers Program, which utilizes state university resources to provide technical assistance to local business, industries, and local communities. There are more than 30 university centers across the United States. Each center is jointly funded by EDA and the university with the expectation that within typically 5 years the center will be fully supported by the university. In 1977 an expanded operational concept for the University Center Program was tested in the southeastern states in cooperation with NTIS and laboratory members of the Federal Laboratory Consortium for Technology Transfer. The underlying concept was to supplement the university resources with the sizeable science and technology base of the federal government residing within NTIS and the consortium laboratories. The one year demonstration project was initiated in 1977 and focused on the problems and needs of small- and medium-sized enterprises. The combination of "passive" NTIS data banks and "active" personal contacts by personnel in the universities and federal R&D laboratories was considered to be an optimum mix for effective technology transfer. Additionally, NTIS also initiated a personal referral service as part of the demonstration project to link telephoned inquiries from the universities or small businesses to appropriate specialists within the federal establishment. It was recognized that lessons learned from the demonstration project would constitute a useful foundation for building an eventual national technology delivery system.

The welfare of small businesses in the United States generally falls under the jurisdiction of the Small Business Administration, an independent federal agency created in 1953 to provide assistance to small business concerns and to help them obtain a fair share of government business. The SBA initiated a Technology Application Program in 1967 specifically to provide technical assistance to small businesses and to help them benefit from the government's investments in R&D. Because the SBA does not conduct its own R&D programs, the Technology Assistance Program therefore is designed to link small businesses to technologies developed by other federal agencies. Major corporations are also not overlooked as potential sources of technology for small business manufacturers or R&D enterprises. Technology having potential industrial or commercial applications are published and distributed to a large mailing list of small business concerns. To supplement the publication and distribution of technical literature, a field network of technical assistance officers is maintained covering all 10 federal administrative regions of the United States. These officers establish personal contact with the small businessmen, who typically were prompted to inquire by the descriptive literature they received. They help the small businessmen define their technical problems or needs and then attempt to locate the appropriate technical data or solution. The ultimate goal is to improve the competitive position of small companies in the domestic and international markets through the use of more efficient production processes or the development and marketing of new or improved products and services. Small business R&D firms also stand to benefit from the technical assistance rendered in being able to make more definitive R&D plans or to expedite a particular R&D project.

The NSF, primarily engaged in fostering basic and applied research, nevertheless supports two novel programs of particular note designed to accelerate the innovation process in the private sector. The first one, called Small Business Innovation Applied to National Needs, is a three-phase program to develop fuller utilization of small, innovative businesses in federal R&D activity and to facilitate the flow of their R&D results to the commercial market.[23] The basic tenet of the program recognizes the exceptional innovative nature of small, technology-based firms and attempts to couple this asset with the larger financial and production base characteristic of large companies. The coupling is achieved via an infusion of outside risk capital from joint-venture partners or venture capital firms. The program expressly solicits high risk, potentially high-payoff proposals from small business concerns. Phase 1 addresses the feasibility and practicability of the research idea and is funded at about $25,000 for six months. Phase 2 is awarded if feasibility and practicality are established and if a venture capital commitment has been obtained by the small business firm. After the more comprehensive second-phase R&D effort is successfully completed, commercialization proceeds under Phase 3 with private capital and public funds to the extent that federal objectives continue to be met. The major attraction clearly is the financial incentives offered by the program that makes available front-end, high-risk R&D funds and lowers the attendant risk for downstream private capital needed for commercial development. The program therefore addresses a major obstacle, that of capital availability, in bringing new technology-based ideas to commercial fruition.

The other NSF program of particular note is the five–year experiment started in 1973 to test the use of innovation centers as a step in improving the climate for innovation in the United States.[24] Centers were established at the Massachusetts Institute of Technology, Carnegie-Mellon University, and the University of Oregon. The prime objective is to determine whether classroom instruction in engineering, business, and technological innovation could effectively foster increased entrepreneurial activity and accelerate the introduction of new products and services into the commercial market. The instructional material emphasized the different stages of the technological innovation process: idea generation, evaluation, product development, business planning, and marketing. New ideas may be generated within the innovation centers or by outside individuals or organizations. Aside from the instructional and training aspects, the innovation centers assist students and participants in setting up new ventures and in getting the needed financing for the project. The innovation center generally retains equity participation in the new venture. Profits from the new ventures accrue to the owners with the university receiving a percentage in relation to its equity interest. Income generated from these new ventures, royalty revenue from the licensing of new products, grants from public and private foundations, and continuing support from the university and industrial concerns are expected to be sufficient to make the innovation centers substantially self-sustaining after the fifth year. The federal government expects to gain considerably from the higher tax base arising from profits of the new companies and wages from the new jobs created. In 1976 alone, the federal taxes collected from company profits and employee wages amounted to nine times the annual federal investment in the innovation centers program. The hope is that the increased entrepreneurial activity fostered by the innovation centers will contribute significantly to a higher technological innovation rate in the United States.

The brief sketch of the activities of different federal mission agencies given here by no means covers the breadth and scope of federal government involvement in technology transfer and innovation. Other agencies with substantive programs include the Departments of Energy, Transportation, and Interior as well as the Environmental Protection Agency. This overview, however, is indicative of the wide recognition pervading the government of the need to make better use of federal technological resources. The many diverse programs reflect the unique mission responsibilities of the federal agencies. In many cases the nature of technology transfer activities are dictated by the need to meet mission responsibilities. In cases where technology transfer is not expressly mandated, the level of effort and allocated resources are subject to and constrained by higher mission priorities of the federal agency. What is evident though is the minimal amount of coordination among the diverse agency programs and activities. No integrative mechanism exists at the federal agency level for steering and coordinating technology transfer activities. A modicum of coordination is achieved, of course, at the R&D laboratory level via the Federal Laboratory Consortium for Technology Transfer, but not all federal agencies have R&D laboratories and, of those that do, not all are participating in the consortium.

The Federal Laboratory Consortium is notable in this regard in that it provides an informal forum for interagency information exchange. It was originally formed not by executive or congressional dictum but simply through the common interests of several DOD laboratories. Leadership in the formation of the DOD Laboratory Consortium was exerted by a major Navy R&D center with the encouragement and cooperation of the Office of the Director of Defense Research and Engineering, with the able assistance of a private consultant.[25] The principal objective is to foster greater utilization of federal R&D resources in dealing with domestic problems affecting the socioeconomic well-being of the nation's citizenry. The formation and subsequent rapid growth of the early, loose association of DOD laboratories into the present Federal Laboratory Consortium with over 180 member organizations was considerably aided and abetted by the widespread recognition that the nation's technological resources lie underutilized and that a sizeable proportion of the nation's technology resides in the federal R&D centers and laboratories. Despite its large size, the consortium was careful not to "bureaucratize" the organization but instead emphasized person-to-person contact, which is generally considered both an important and necessary ingredient for successful technology transfer. The consortium works through technology transfer representatives at each of the member laboratories who are so designated by their respective laboratory managements. In the early 1970s the focus of activity was in interagency technology transfer, that is, in facilitating the flow of technology between federal agencies where it may best address domestic priorities in areas such as health, transportation, environmental pollution, and energy. During the mid 1970s attention was increasingly shifted to helping solve state and local government problems. The major impetus in this direction was provided by the National Science Foundation inasmuch as a consortium representative had been assigned to the intergovernmental science area of NSF in the early days when the consortium was first formed. The intergovernmental science function is specifically aimed at bringing the benefits of science and technology to states and municipalities. The location of the consortium representative within the intergovernmental science area naturally led to efforts to bring federal laboratories' technology to bear on meeting the needs of state and local governments. Little attention was given to transferring technology to the private sector until a consortium representative from a Navy R&D center was assigned temporarily in 1976 to NTIS within the Department of Commerce expressly to work towards the commercialization of the federal laboratories' technology. The expanded concept for the Economic Development Administration's university center program was consequently developed as a means by which technology in the federal laboratories could emerge in the form of new and improved products and services in the commercial sector. Consortium orientation towards developing industrial and commercial uses of technology in the federal laboratories has always been inhibited by uncertainty as to how to go about it. Technology transfer to industry requires more than the customary client-contractor relationship. However, closer working relationships are constrained by the necessity to avoid situations that might be construed as conflicts of interest or violations of government impartiality towards private-sector organizations. Still, the indisputable fact remains that federal technology transfer, if it is going to contribute in any significant way to spurring the nation's productivity growth and industrial competitiveness, must be concerned in a major way with strengthening the technological innovation process in the private sector.

Yet the limitation of the interpersonal approach lies in its narrow scope as compared to the pervasiveness of government statutes and regulations. If through consortium activities the dedication of a handful of technology transfer activists in the laboratories could be made to spread to other member organizations, the effectiveness of interpersonal efforts will multiply many fold. It must also be recognized that the technology transfer function is not in the mainstream of business at many of the laboratories but is merely one of many secondary collateral activities. The consortium therefore serves significantly as a mutual reinforcement mechanism for laboratory representatives engaged in technology transfer as well as a legitimizing outlet for nonmission-oriented work in the federal R&D centers. However, the most important function of the consortium is in being able to aggregate the collective resources of America's vast network of federal R&D centers and laboratories and to present these resources in a highly visible manner to the user communities in the public and private sectors. It serves to build a higher awareness in the user communities of the availability of federal technological resources and acts to make the science and technology base more easily accessible to potential users. Yet for all the altruistic efforts of the federal laboratories in promoting the transfer and utilization of their respective technologies, much depends on the inclination and disposition of the potential user to want to seek and use federal technology.

A growing movement does seem to be developing in state and local governments to seek information from the science and technology community. Attendant with this trend are joint federal and state efforts to strengthen state capabilities for utilizing science and technology. Many states have designated a state science advisor. State legislatures are recruiting for public policy scientists to interface between science and law. At the local level technology agents are working to improve community services that are increasingly dependent on technological solutions. Cities and counties are organizing into national and regional networks with federal assistance to aggregate collective needs and to pool resources. Many of these networks are linked to the Federal Laboratory Consortium for technical assistance. High priority is given to addressing needs and problems common to many municipalities. Attention also is given to reducing the typical mismatch between municipal problems and the leading-edge technologies found in the federal R&D establishments. Though much progress can be cited in the growing receptiveness and ability of states and municipalities to utilize science and technology, considerable room exists for further innovative ideas and approaches.

One innovative approach pioneered in Connecticut is the Connecticut Product Development Corporation (CPDC), a state agency set up to create new jobs within the state by providing financial assistance to small businesses for new product development.[26] The basic premise is that new products help companies grow and help expand employment opportunities in the state. The agency was started with proceeds from the issuance of state general obligation bonds to be eventually repaid with interest. CPDC by law must become self-sustaining from income derived from new product ventures that it helps finance. Its basic operation is to provide risk capital to small companies specifically for the purpose of developing new products. Small companies are emphasized because they, more than large companies, experience great difficulty in obtaining capital for risky new product ventures. Traditional lending sources rarely will finance new product developments in small companies. As a result industrial expansion and job creation are stymied. CPDC, however, will provide risk capital for new product development without requiring equity participation in the venture. Moreover, the risk capital is not provided as a loan that appears as a liability on the company balance sheet. Basically, CPDC enters into a joint venture with the company solely for developing the new product. Typically, CPDC provides 60 percent of the capital and the company advances the remaining 40 percent. A 5 percent royalty on eventual product sales is retained by CPDC until it recoups five times its original investment, after which the royalty rate drops to half a percent. If the new product development fails to reach commercialization, CPDC stands to lose part or all its investment. In this case the company is not obligated to repay the funds. Based on operational results since 1975, its first full year of operation, CPDC has experienced a success rate of over 80 percent on the ventures it approves for financing. Royalty income derived from new product sales is rapidly expanding so that self-sustaining operation promises to be achievable within the near term.

CPDC is the first of its kind in the United States. It was patterned after the National Research and Development Corporation in the United Kingdom, but was modified to fit Connecticut's special circumstances. Perhaps the closest similarity to CPDC is found in Massachusetts, where the Massachusetts Technical Development Authority also extends financial aid mainly to inventors in order to help attract additional venture capital from private sources.[27] More and more states are inquiring about CPDC and looking to ascertain whether a similar organization would be appropriate in their states. Although an organization similar to CPDC would not be suitable in all states, it nevertheless promises to be an effective instrument for stimulating economic development and job creation in many states where small business development is a major economic factor.

The diverse activities briefly described here are not exhaustive but are indicative of the national concern for revitalizing technological innovation in the United States. This concern is by no means confined only to the public sector. During the 1970s more and more of the technical societies, trade associations, and professional organizations have formed working groups and committees and organized conferences to consider various aspects of the overall subject. Evidence of the new technology consciousness can be found by simply scanning the names of the new magazines and journals recently introduced—*Journal of Technology Transfer, Technovation,*

Planned Innovation, and *Technology and Society*—to name just a few. Additionally, a new international society for technology transfer was formed to provide a forum for deliberations and mutual sharing of experiences on the subject. These developments reflect a genuine consciousness and concern for the nation's socioeconomic problems and are indicative of the widespread recognition that in technology lies the key to curing many of our socioeconomic ills.

Fiscal and Regulatory Incentives

United States tax policy is one area that many agree can and should be changed to stimulate innovation. Inadequate incentives exist in the current tax code to encourage capital formation and investment in new technology-based enterprises. Provisions of the tax code now permit R&D expenditures to be considered as operational expenses during the year incurred or capital assets to be amortized over a period of not less than five years. Investments in technological plant and equipment are subject to the normal depreciation schedules as for other business capital assets. Businessmen contend that the lack of preferential tax treatment for R&D activity and accelerated depreciation of associated plant and equipment deter capital accumulation and investments in productivity-enhancing innovations and new entrepreneurial ventures.[28] Adequate fiscal incentives are needed to cushion the added financial risks attendant on technology-based innovative activity. Especially in an era of high inflation and expensive money, investment capital tends to migrate toward safer, less venturesome havens. This is evidenced by the scarcity of venture capital in the mid 1970s, which made it extremely difficult for entrepreneurs and promising young companies to raise needed capital during that period. Only after the maximum capital gains tax rate was reduced from 49 to 28 percent in 1978 did venture capital begin a comeback. The important consequence of this single change in the tax code illustrates the powerful potential of using tax incentives as a means to encourage new ventures and innovative activity.

Preferential tax treatment to encourage technological innovation by no means infers irreparable loss of government tax revenues. On the contrary, the increased economic activity from new products and new companies established from increased availability of venture capital produce a growing tax base that will help to mitigate the revenue loss from more liberal tax incentives for R&D expenditures and new ventures. Indeed, the federal government is in fact a majority partner and beneficial owner of American business through its tax share of business revenues. A major portion of business income flows to government coffers in the form of corporate taxes and individual income taxes imposed on wages and declared dividends. Therefore, the public interest is also served by providing more attractive tax incentives to spur capital formation for innovative new ventures and entrepreneurial activity. Although widespread agreement exists on this score, the major question is whether agreement can be reached on the exact nature of the tax revisions to be implemented. Some of the changes considered include faster depreciation schedules for buildings and R&D equipment, a 10 percent value-added tax that encourages individual savings and investments at the expense of consumption, elimination of double taxation of corporate dividends (via corporate and individual income taxes), tax credits for contributions to research institutions, and deductions for investments in new technology-based enterprises.

Greater capital accumulation and investments in new, innovative ventures can be encouraged not only through added tax incentives but also by reducing the amount of excessive or poorly conceived government regulatory requirements. Procedures necessary to comply with government regulations and their attendant reporting requirements increase the cost of doing business. For large and small companies alike the cost for meeting regulatory standards correspondingly reduces the amount of capital available for plant modernization, new product development, and investments in new ventures. The reward-to-risk ratio is lowered, driving available venture capital to seek other less-regulated investments with lower levels of risk and surer, albeit lower, rates of return. The impact of diminished access to investment and venture capital hits the small business concerns the hardest as they are more heavily dependent on equity than are the large companies. Because the small businesses must necessarily be more innovative to stay competitive, constricted access to the capital markets will have a detrimental effect on the nation's innovative ability. Proposed changes in the regulatory climate affecting small businesses are under consideration, such as placing the burden of proof on the regulatory agency to establish a cause for concern before requiring regulatory compliance by a small business and allowing small businesses a double tax deduction for expenses incurred in obtaining regulatory advisory services related to compliance with government regulations.[14] Furthermore, relaxation of the Securities and Exchange Commission rules for registration of public offerings of stock from small companies has been advocated so that small businesses could gain needed equity financing more easily and in a more timely fashion.

Another regulatory area significantly affecting the climate for innovation concerns U.S. patent policy (or, more aptly, policies inasmuch as a uniform federal patent policy does not exist). The U.S. Constitution empowers Congress to "promote the progress of science and useful arts, by securing for limited times to authors and inventors the exclusive right to their respective writings and discoveries." Thus the patent system is intended to provide the incentive of legal protection to an inventor in order to encourage public disclosure of his invention to speed technological progress and the invention's adoption in commerce. In return for full public disclosure, the inventor receives a 17 year exclusive right to develop, use and sell his invention. However, over the years Congress has acted in a fragmented, inconsistent fashion.[6] As a result, numerous policies have evolved that differ from agency to agency dealing with rights of the government and the inventor in cases where inventions are made as a result of federally sponsored R&D. Furthermore, in those federal agencies where their statutory authority governing patent rights are unclear, they must rely on interpretations of the Presidential Memorandum on Government Patent Policy issued by President Kennedy in 1963 and subsequently revised by President Nixon in 1971. This has consequently led to nonuniform interpretations and determinations of inventor rights depending on the particular federal agency concerned. Nonuniformity of federal patent policies also extends to the domain of privately developed inventions and patents. Consequently, federal patent policy inconsistencies and issues pertaining to the equitable distribution of invention rights that would best serve the public interest act more as disincentives to inventors and are therefore contradictory to the original intent of the constitutional authority granted Congress.

A major issue since federal patent policy was first enunciated in 1943 has been whether title to inventions resulting from government-sponsored R&D should reside with the government or the inventor-contractor with government acquiring a royalty-free, nonexclusive license. Proponents for government title retention argue that because the invention was made with public funds, it should therefore remain in the public domain. Opponents, on the other hand, contend that government patent ownership defeats the original intent of patent policy, which is to promote full disclosure and rapid utilization in the public interest. A government patent freely available to all offers no protection to the potential licensee who may have to devote large sums of money and considerable effort to commercialize the invention. It is argued that only when the title remains with the inventor-contractor with government retaining royalty-free, nonexclusive license will the necessary incentive exist for the contractor to pursue commercial exploitation of the invention.

The government over the years has by no means been of one mind on the title issue. While early government pronouncements provided for government retention of title, often referred to as the "title policy" concept, various federal agencies registered opposition to the policy and instead supported a "license policy" where government retains royalty-free, nonexclusive license and title remains with the contractor. This coupled with the notably different missions and statutory responsibilities of the various departments and agencies has contributed to the nonuniform patent policies presently in effect. Although the title policy was in early favor with most agencies, the current trend is towards a more flexible government posture that embraces both title-policy and license-policy concepts.[4] The Presidential Memorandum of 1963 as revised in 1971 enlarged the authority of agency heads to waive the government's title to inventions in favor of the contractor and to permit exclusive licensing of government patents. Exclusive licensing is permitted when, after publication for at least six months, use of the invention has not been achieved under a nonexclusive license. The public is protected by strong "march-in" rights under which the government may require licensing of the invention to a third party if it is deemed in the public interest or if the contractor has not made sufficient progress in commercializing the invention. The increased flexibility embracing the license-policy concept reflects the growing recognition that commercialization of inventions is more likely to occur when the title resides with the contractor rather than with the government. This view has become more prominent amid mounting evidence that the government's title policy has not provided a sufficiently strong incentive for a respectable level of inventive activity stemming from government R&D sponsorship, nor has it served to encourage private-sector licensing of government-owned patents. Though the government supports roughly half of all R&D in the United States, private-sector patents outnumber patents derived from government-supported R&D by a factor of better than 20 to 1.[5] In addition, despite the fact that about two thirds of government R&D is contracted to the private sector, contractor inventions account for only about 40 percent of government patent applications.[29] Also, less than 5 percent of the government patents are licensed by industry. Although many factors contribute to the low patenting activity from government R&D and the low licensing rate of government patents, a major factor undoubtedly is the small incentive provided by the government's title policy. As a result business and industry prefer to use their own resources in performing significant R&D that may lead to important inventions rather than to rely on the vagaries of government R&D sponsorship and patent policies.

Another circumstance that detracts from the desirability of government-owned patents is the laxity of most federal agencies in obtaining foreign patent protection. With the exception of NASA, DOD, HEW, and the old AEC, federal agencies have generally ignored the foreign commercial potential of their inventions which by nature should be readily suited to commercial applications.[30] As a result foreign manufacturers are able to use U.S. patented technology without paying royalties. Equally serious, domestic licensees of government patents are not provided with foreign patent protection, which would be important if they wished to exploit the government invention overseas. Generally, the federal government has six months after the U.S. filing to exercise its option to file patents in foreign countries on government-owned inventions. If the six month period lapses without action by the government, foreign patent rights revert to the inventor. However, most foreign countries bar patent filings if the invention has been previously published, unless the U.S. patent application predates the publication, in which case foreign filing can be made within one year of the U.S. filing date. Therefore, in order to preserve foreign filing rights, government inventors should see that the U.S. patent application predates any publication of their invention.

To provide stronger foreign patent protection for government inventions, the Secretary of Commerce in 1950 was granted authority by executive mandate to receive custody of foreign rights to government inventions from other federal agencies, to seek foreign patent protection, and to license these inventions. Subsequently, the NTIS has pursued an active foreign-filing program in order to provide foreign protection for U.S. licensees of government inventions and to obtain a return for the federal government on U.S. inventions used abroad. Under this program NTIS works with other federal agencies to acquire foreign rights to selected government inventions and files for patent protection in foreign countries. After foreign patents are filed, NTIS undertakes to license them either to American companies seeking to practice the invention abroad or to foreign enterprises in the event no American licensees were found. This program is supported with appropriated funds, and royalties generated from the licensing of foreign patents are returned to the U.S. Treasury. In 1977 NTIS was given additional authority to inaugurate an incentive awards system for government inventors whereby a small percentage of royalties realized from licensed inventions are returned to the government inventor. This awards system provides added incentive for the government inventor to seek foreign patent protection through NTIS.

Privately developed inventions and patents are also subject to various federal patent policies particularly when they are "background patents," that is, private-sector patents that form the basis for subsequent inventions made under government contracts. Although the policies of the federal agencies differ in this respect, many agencies and departments require the compulsory or mandatory licensing of background patents insofar as such licensing is necessary to carry out the government contract work. Not surprisingly, this mandatory licensing aspect is a serious concern to companies and serves as a disincentive to competition and private investment in R&D. Generally, however, the federal agencies have acted in good faith in recognizing the rights of the owner, requiring mandatory licensing only when in the public interest and providing just compensation to the patent holder.

Other patent issues exist, such as reconciling with antitrust laws the limited monopoly conveyed by patents and strengthening the patent search process[30] inasmuch as an estimated 80 percent of patents challenged in federal courts have been ruled invalid.[31] Efforts have been under way to address these issues and to establish a uniform federal patent policy relating to inventions resulting from government-sponsored work. Provisions of a possible new federal patent policy would include the allocation of rights to inventions resulting from government programs, protection of these rights through patent filings in the United States and abroad, and licensing and commercialization of the invention technology. The trend is toward the license-policy concept provided there is strong intent to pursue commercial exploitation by the contractor, but the federal government would reserve strong march-in rights to protect the public interest if deemed necessary. There is therefore some reason to anticipate in the near future the establishment of a federal patent policy more conducive to technological innovation in the United States.

CHAPTER 7

Meeting the Challenges Ahead

It is not surprising to find that the climate for innovation in the United States, West Germany, and Japan is distinctly different. All three nations are world economic powers, but only the United States has experienced substantive erosion in productivity growth and international competitiveness in recent years. The obvious question is "Why the disparity in American economic performance compared to that of West Germany and Japan?" What are the principal factors that underlie the adverse economic trends in the United States compared to the other two countries? In attempting to get the answers, it would be useful to identify the major institutional and policy differences affecting technological innovation in the United States vis-à-vis West Germany and Japan. Specifically, we should try to pick out those factors that share commonality in West Germany and Japan but differ from conditions found in the United States. Although specific foreign experiences may not be readily adaptable to the American scene, they nevertheless could provide the basis for formulating additional, perhaps also more enlightened, policy options. Such an evaluation should in turn lead to more informed judgments as to the major institutional and policy changes needed to help restore America's economic vitality.

Comparative Assessment of the United States with West Germany and Japan

In comparing the climate for innovation in West Germany and Japan, nine points of commonality were established that could reasonably be argued to be the significant determinants of economic progress in the two countries. They represent the common threads that make up the underlying fabric of

197

the German and Japanese postwar economic resurgence. By considering each of the nine points of commonality in light of corresponding conditions in the United States, we can ascertain where significant differences exist, the implication being that an analysis of the differences can suggest appropriate avenues for improving the climate for innovation in the United States.

Looking first at the character of American workers compared to German and Japanese workers, the Germans and Japanese are found to be more frugal as evidenced by their higher savings rate compared to American workers. German workers manage to save three times as much of disposable income as American workers. Japanese workers on average save about four times as much of disposable income than American workers. A number of factors account for the comparatively lower savings rate in America. The higher inflation rate in the 1970s has spawned in consumers a buy-now mentality in anticipation of further inflationary price increases. The decline in the savings rate is also attributed to consumer perception that there is less need to save for retirement or for unforeseen "rainy days" in view of the many social benefits available such as social security, retirement and pension plans, Medicare, and government aid to the poor and disadvantaged. Furthermore, the decades of American prosperity have given rise to greater confidence in the ability of the national economy to rebound from adversity. Moreover, unlike foreign cultures, the incurrence of debt is virtually a way of life in America. On the other hand, German and Japanese workers rely less on credit purchases than their counterparts in America. As a result, the higher savings rate in Germany and Japan makes more money available for financing industrial growth and at the same time helps to limit inflationary pressures in their respective economies.

The Germans and Japanese are also known to excel in worker industriousness. The stronger German and Japanese dedication to hard work is manifest in the appreciably higher worker productivity gains they achieved as compared to those experienced in the United States. Meanwhile, in America changing social attitudes have led to stronger desires for the "good life." A softening of the American work ethic is reflected in worker demands for fewer working hours and higher pay. A corollary is the shifting of the American economy from an industrial to a service economy. Approximately 60 percent of the American work force is engaged in service activities including government service. This change in the nature of the American economy has also been cited as a reason for lowered productivity growth inasmuch as productivity in the service sector is believed to rise slower than industrial productivity. Another important factor in lowered U.S. worker productivity is the greater disposition of the American worker to go on strike for higher wages, benefits, and better working conditions. The man-hours lost in work stoppages are irretrievable and constitute a costly interlude for both the company and its workers.

This points to a marked difference between the nature of labor-management relationships in America and the nature of labor-management relations in West Germany and Japan. The adversarial attitudes of confrontation that typify the relationship between unions and managements in America are notably absent in West Germany and Japan. Labor-management relations in Germany and Japan are founded on developing mutual consensus through greater worker participation in management affairs. The views of workers are channeled to management through numerous committees and working groups. Worker representation in management councils is more the rule than the exception. In contrast, the demarcation between management and labor in America is much sharper. Workers are typically not involved in the management of their company. When Chrysler first announced in 1979 its intention to nominate the president of the United Auto Workers to its board of directors, it represented the first such move for any major corporation in the United States. The announcement was received with widespread scepticism and even some opposition from the union ranks. Criticism was voiced that placing a union representative on a corporate board of directors constitutes a conflict of interest. This critical reaction signifies the depths of labor-management division existing in America and overlooks the obvious convergence of interests in that both labor and management stand to benefit when the company prospers. In fact, having worker representation on corporate boards of directors is in widespread practice in West Germany.

In relationships between government and industry, both West Germany and Japan exhibit closer cooperation and provide greater governmental assistance despite the fact that their commitment to the free-market system is no less than that found in the United States. The German and Japanese governments are closely attentive to the needs of their respective industries and extend considerable assistance to encourage innovation and to strengthen industrial competitiveness. Forms of assistance include forming joint ventures on industrial projects, increasing the availability of risk capital, providing market research activities, and attending to the specialized needs of small- and medium-sized companies. However, government largess does not extend to keeping noncompetitive firms alive. On the other hand, the relationship between government and industry in the United States is dictated typically by the formal contract between customer and contractor. The fortunes of private enterprise are by and large determined by the vagaries of market forces associated with the free-enterprise economy. Government measures to stimulate innovation and industrial competitiveness are not nearly as extensive as those found in West Germany in Japan. On the contrary, recent U.S. Government interventions in the market economy have taken the form of increased regulatory requirements that are widely viewed as strong disincentives to free enterprise in America. More government attention is given to protecting and preserving industrial sectors that are experiencing greater inability to cope with foreign competition than is devoted to restructuring and concentrating resources in areas where competitive advantages exist and can be sustained. Moreover, government efforts to stimulate industrial innovation are characterized by appreciably divergent views as to the appropriate policy measures needed.

Both West Germany and Japan recognize that R&D constitutes an important instrument for industrial progress and economic growth. The major share of R&D is funded and performed by the private sector. Industrially relevant R&D is heavily emphasized. Even in government-supported R&D projects, much attention is given to developing industrial and commercial applications. In the United States generally no more than about half of R&D expenditures comes from industry. Government expenditures for R&D, moreover, have been heavily concentrated in the military, aerospace, and nuclear energy categories. Only in recent years has the share of civilian-oriented R&D risen in fields such as transportation, environmental protection, and alternative energy sources. However, it must also be recognized that even greater investments in civilian R&D do not necessarily translate directly into industrial and commercial applications. Furthermore, R&D programs in the United States are decentralized and organized along mission-agency lines. Financing and performance of industrial R&D are generally left up to private enterprise, though government financing in the manufacturing technology area is a notable exception. However, because of the slowing innovation rate amidst declining economic indicators, stronger government measures for spurring technological innovation are being deliberated. Progress has been slow due to the unprecedented nature of greater government involvement in areas traditionally reserved for private enterprise. Despite the absence of a national science and technology policy and a clear consensus as to needed government measures to revitalize industrial innovation, there is a growing movement to transfer technology and to develop greater utilization of the science and technology base.

A closer look at the pattern of technology transfer in West Germany and Japan shows some marked differences with the situation in the United States. Whereas technology-intensive products account for the major share of export trade in all three countries, on balance the flow of technology is inward in both West Germany and Japan and outward in the United States.[1] That is, exports of technology-intensive products in Germany and Japan is balanced in part with a net inflow of technological know-how from foreign sources. The inflow of technology could in fact be regarded as helping to fuel the German and Japanese industrial engine and their export trade in technology-intensive products. On the other hand, the net flow of both technological know-how and technology-intensive products in the United States is outward. The implication is that unless a more balanced flow of technology is developed in the United States in the form of greater technology feedback from abroad, America's technological advantages and surplus trade in technological exports cannot be sustained in the long term.

West Germany and Japan also have linker organizations, such as the Garching Instrument Company in Germany and the JITA in Japan, established to help bridge the gap between research and industrial applications. These organizations specialize in tracking technological developments at home and abroad and seek to match these developments with needs of the industrial sector. No comparable organizations exist in the United States in spite of the rising level of technology transfer activity. The closest semblance to linker-type activities are provided by the independent consultants who would normally contract to provide similar services on specific projects. In addition, government technology transfer agents also perform many of the linker activities. However, there has not been any government or quasi-government organization set up specifically to expedite the industrial and commercial applications of research results in the United States.

Certainly, the system of tax and regulatory incentives prevailing in the United States is not as extensive as that found in West Germany or Japan. The incentives employed in Germany and Japan more broadly and directly address the different stages of the technological innovation process. A broad array of tax incentives are employed in both countries to encourage R&D activities as well as the capital investment required to translate technological progress into industrial and economic growth. Both countries also provide more extensive incentives to encourage inventive activity than the United States. For example, in Japan government inventors whose patents are licensed to private enterprises by the JRDC are awarded 90 percent of the royalties that JRDC receives.[2] Likewise, West Germany uses tax and patent measures to spur the creative activity underlying the technological innovation process. The Germans also place considerable emphasis on protecting the rights of the inventor. Japan has also not neglected the market and informational aspects of the innovation process. In particular, Japan has established tax benefits for promoting technological exports and has set up organizations devoted to collecting and disseminating technical and market information of use to Japanese industry. The United States, by comparison, is sorely deficient in extensive fiscal and regulatory incentives to foster technological innovation. Available tax benefits generally center on conventional write-off of business expenditures and allowance of limited investment tax credits with virtually no provisions for incentivizing other stages of the innovation process. Undoubtedly, this situation has contributed in a major way to the recent slowing of the innovation rate in the United States.

Small- and medium-sized enterprises are given special consideration in Japan, West Germany, and, to a lesser extent, in the United States. In Japan the smaller firms are entitled to more favorable tax benefits than the large companies in the form of higher tax credits for R&D, higher depreciation for capital equipment, lower corporate tax rate, and tax exemptions on local taxes. The Germans, however, focus on developing better access by small- and medium-sized companies to sources of risk capital through financial credits and guarantees to investment houses and venture capital companies. Additionally, some German state agencies provide technical assistance to small companies to enhance their ability to innovate.[3] In the United States, special treatment for small businesses generally involves giving added consideration to them for government procurements, contracts, and loans. More recently, a larger share of federal R&D funds is being set aside for small business awards. Also, limited technical assistance to small- and medium-sized companies is being provided through the Technology Assistance Program of the Small Business Administration. However, the extent of available tax incentives is not comparable to that available in Japan, nor is access to sources of risk-capital made easier by means of government financial assistance as in West Germany. For the small business enterprise, capital availability and financial credits are critical factors in its ability to pursue new product innovations. A number of legislative proposals for providing greater assistance to small businesses have already been introduced in Congress but, unless and until they are enacted, the small-business sector will continue to experience difficulty in the capital markets.

Implications for Institutional Change

Observers of the American scene can undoubtedly discern a strengthening in America's resolve to try to correct the imbalances that contributed to the economic malaise afflicting the nation in the 1970s. Politicians and pundits alike have proffered their views and recommendations on what needs to be done to set the country right again. It must be recognized, however, that the needed institutional changes involve more than simply starting new and costly government programs; they reach out to the very core and foundations of the democratic form of government and the free-market economic system. How to accommodate the needed institutional changes and yet leave inviolate the principles of free enterprise and democratic government constitutes the major challenge. To what extent can greater government intervention and assistance in the free-market economy be permitted without incurring the risk that a form of creeping socialism will set in? Yet it has always been the responsibility of management, no matter whether in government or industry, to manage change in a way that established goals are reinforced and adverse effects softened. This is even more true today when the United States must increasingly consider various forms of governmental intervention in the free-market economy.

Traditional American economic philosophy has always been to preserve the unconstrained operation of the free-market economy. Reliance was placed on prevailing free-market forces to correct temporary imbalances in the marketplace. Industries will wax or wane in accordance with the dictates of market forces. This philosophy is becoming increasingly challenged as one that represents an abdication of government leadership in setting the tone and conditions for national economic prosperity. One must consider also the much longer times it now takes for market forces to reach an equilibrium state given the complexity of modern society. As a result the human, social, and political consequences of market-economy imbalances are considerably magnified from the time of a more simpler age. Society is becoming less tolerant of economic adversity.[4] This is not to say that the free-enterprise system should be replaced with a socialistic one. American commitment to the free market remains unshaken. It does, however, augur for greater governmental leadership in guiding and creating the appropriate conditions for a thriving free-market economy. Industrial and trade policies must be clearly formulated to permit appropriate resource allocations and industrial restructuring that would be anticipatory rather than reactive to market forces. Government intervention is needed to accentuate long-term trends that coincide with established goals and to ameliorate human readjustment problems when economic forces run counter to desired trends and goals.

There should be full recognition of the fact that competitive conditions are dynamic and constantly changing. Industries that are competitive in domestic and international markets today will not necessarily be so tomorrow. As the industrial challenge from the developing nations mounts, U.S. industry must continually upgrade and move into newer areas where superior technology and production know-how can provide the needed competitive advantages to sustain a mutually beneficial, complementary trade relationship with the developing nations. Investments in emerging technologies and innovations that could form the basis for future new, competitive industries should be emphasized. Continual efforts must be made to distinguish industries that are competitive in world markets against those that are noncompetitive. Competitive industries should be nurtured and supported through stimulative fiscal, monetary, and regulatory policies. Noncompetitive industries should be encouraged to redirect resources into other more promising areas or allowed to wither. Assistance should be directed towards phasing out noncompetitive businesses and to soften the human readjustment problem through retraining and relocation. Resources should, however, not be concentrated in defending weak, noncompetitive industries but should favor the strong industries that can continue to be competitive as well as the emerging growth industries that will be the principal source of new jobs in the future. However, the record shows that much political capital continues to be expended on protecting our less competitive industries, such as steel, textile, and shoes, from the foreign challenge. Additionally, huge sums of public funds are being committed to stave off bankruptcies for major corporations such as Lockheed and Chrysler in order to avoid throwing large numbers of people out of work. Though preserving jobs is important, the jobs must nevertheless be productive and in competitive industries. Proper emphasis should be placed on fostering future competitiveness rather than preserving outmoded institutions.

The greater complexity of modern society and government institutions also gives rise to more complex public issues that require resolution times appreciably longer than those required in a simpler era. Where major public decisions could be implemented within months, now many years is more typical. As the time period required to introduce changes progressively lengthens, a marked incompatibility develops between the shorter time horizons important to political leaders and industrial managers and the longer time required to resolve complex issues and problems. If attention to major policy issues will not provide timely dividends by the time of the next election, the politician will naturally expend his energies in other directions. Similarly, the industrial manager will tend to concentrate his efforts in areas that will produce improved near-term operating results rather than focusing on high-risk, innovative ventures that will come to fruition, if at all, only after many years. As a result long-term policies and solutions do not attract the keen attention of key figures in the public and private sectors. Even if in some cases vigorous pursuit of long-term solutions is made, there is nevertheless the question of continuity of leadership when personnel turnover in the United States occur typically in a two to four year time frame. Public issues are sufficiently complex and diverse that even the brightest political minds will not have the staying power or corporate memory required for doggedly pursuing and resolving the many details and side issues as well as for implementing appropriate institutional policies in response. This incompatibility in the time horizons of the nation's leadership and the demands of society and government consequently detracts from our collective ability to deal with national problems and issues in a timely and effective manner.

Nor is the situation amenable to ready solution. One possible but unlikely alternative is to extend the terms of offices of elected officials. By this means the time horizons of public officials will necessarily be lengthened. However, the attractiveness of this approach is offset by the substantially higher risks associated with less frequent accountability to the public electorate. Longer terms of office could substantially increase the likelihood that real abuses of power will occur. In addition, nothing less than constitutional amendments can alter the terms of offices for many key elected positions. Similarly, longer appointments for industrial managers are no less unlikely. Unless the individual has already developed a glowing record of achievement, few companies would want to be burdened with long-term contracts for employees whose performance falls short of mutual expectations. With these constraining factors, perhaps it is more prudent to consider instead the formation of a small nucleus of able career bureaucrats to provide the long-range direction and continuity of leadership required. Specially trained and seasoned, this elite corps of government managers could be given the authority and oversight responsibility in different areas such as foreign policy, transportation, and energy. The rudiments of such a group already exist in the form of the U.S. Government's Senior Executive Service, composed predominantly of experienced, high-level, senior bureaucrats. However, it is not likely that additional power and oversight authority would be granted career bureaucrats by elected and appointed officials as this move would be tantamount to an abrogation of power, which public officials are loathe to make. Short of institutional changes such as these, the only other alternative would be to appeal to increased individual dedication to tackling long-term problems and issues irrespective of their relationship to personal milestones and horizons.

We must also recognize the limitations inherent in the decentralized, mission-agency structure of the federal government. Without delving into the question of the most appropriate government structure for the United States, it nevertheless is wise to keep foremost in mind the inherent weaknesses of a decentralized form of government—inadequate oversight and coordination—in order to minimize their impact on the governmental process. Mission-agency programs come and go as different waves of popularity surge through Washington, D.C. Often many programs concern two or more different agencies, but narrow mission interests generally do not lead to effective interagency coordination. Furthermore, programs that are interagency in scope rarely enjoy high-level, nonparochial management oversight that spans agency lines. Inadequate oversight and coordination in turn leave unchecked the growth of duplicative activities and facilities among the mission agencies. Moreover, there is always the danger that problems or issues that are not clearly within the mission jurisdiction of a single agency will not be vigorously pursued but instead will languish for lack of proper sponsorship. Unpopular interagency issues may tend to "fall through the cracks" rather than to receive prompt attention. Perhaps the greatest disadvantage lies in the extraordinary effort required to develop cohesive national policies that accommodate the diverse intersts of the various mission agencies and departments. The lack of a national science and technology policy and the absence of a single government patent policy are two prime examples that reflect the decentralized character of the federal government. Interagency coordination is much more likely to occur at the program level than at the policymaking level. Attention to coordinative mechanisms at all levels is therefore of utmost importance. In this vein the concept of a specially trained elite corps of high-level career bureaucrats that could serve as a national pool of management talent assignable to different mission agencies for limited periods of time is appealing from the standpoint of potentially greater interagency coordination arising from the movement of key, high-level personnel between different sectors of government.

Not only would freer movement of people across interagency lines be desirable, but also the increased diffusion of technical information and technologies among agencies would contribute towards more efficient use of available resources and a more stimulative environment for innovation. The disadvantages of a decentralized mission structure could be at least partially compensated by conscious efforts to share problems, knowledge, and resources. In this regard, considerable progress can be detected in the growing federal involvement in technology transfer. Federal technology transfer efforts should be helpful not only in developing greater utilization of technological resources within agencies but also in overcoming interface barriers separating mission agencies. However, technology transfer must be mutually beneficial. It must be rewarding to both source and user alike if the relationship is to endure. For this reason it is important to view technology transfer as fundamentally a two-way process. This bilateral character is equally applicable to all levels of transfer—international, national, and organizational. For the organization more accustomed to being the technology supplier, it is equally important to maintain knowledge of technological progress elsewhere in order that appropriate technology might be acquired to catalyze internal developments to higher and more sophisticated levels. For the technology user more accustomed to purchasing or licensing technology from others, it is equally important to begin building up proprietary knowledge and to strive towards greater technological self-sufficiency and the eventual ability to innovate. A two-way technology flow fosters the mutual interaction needed to stimulate technological progress and the creative process that underlies the ability to innovate. Otherwise, technologies will tend to lie fallow and lose their competitive edge.

Despite the deeper involvement in technology transfer by the federal government during the 1970s, the movement of technology across America's borders is still overly weighted in the outward direction. A more balanced two-way flow is desired so that technologies from abroad could be more effectively used to expedite technological innovations in the United States. The need for a more balanced two-way flow is underscored by the declining R&D trends in the United States coupled with the rapidly developing technological competence of foreign countries. The government can play a major catalytic role in this regard by taking steps to (1) better coordinate and consolidate information gathering activities relating to foreign technological developments, (2) develop appropriate storage, retrieval, and assessment capabilities for effective data management, and (3) assure that foreign technical information resources are available and accessible to strengthen technological innovation in industry as well as to support U.S. policies in areas such as technology export controls and NATO armaments cooperation. At the same time, greater domestic exploitation of foreign technological advances would complement and most likely stimulate fuller utilization of the domestic science and technology base.

Much more needs to be done to promote the diffusion of existing government technologies in the commercial and industrial sectors. Herein lies the promise of substantially higher returns from the public investment in R&D. However, technology in the government domain largely lies underutilized. Because the bulk of federal R&D is contracted to the private sector, a large share of government technology already resides at contractor sites, but much of this technology lies dormant with no attempts made to develop commercial applications. There are many reasons for this state of affairs: (1) the potential commercial market for the technology is unattractive or unsuitable for the contractor, (2) the amount of development capital required for commercialization is too high, (3) the ready availability of the technology to competitors constitutes a strong deterrent, or (4) the commercial potential of the government-owned technology has simply not been recognized by the contractor. Yet it is important to realize that what may appear to be an unattractive market to one company may be very attractive to another. Even if the potential commercial application is unsuited to one company's product line or business plans, it may be directly appropriate to another company in a different industrial context. Where the potential market may be too small for a major corporation, it may be large enough for a small company. Where the capital requirements and in-house know-how needed for commercial production is excessive for one firm, it may not be so for another company better positioned to exploit the technology. Clearly then, what is needed is better dissemination of information about the availability of the government technology throughout the private sector. However, most government contracts do not require the contractor to identify commercially promising technological developments resulting from the contract work. Furthermore, there is no incentive for the contractor to inform other companies about the commercial appeal of available government-owned technology that they have decided not to exploit themselves. The burden therefore rests with the government to establish greater accountability of commercializable technologies resulting from government R&D contracts and to diffuse the information widely within the private sector.

Improved private-sector access to technologies in the government domain should materially spur industrial innovation by fostering new uses of existing technologies in different industrial and commercial settings.[5] Improved dissemination of technical information should also be especially helpful to small- and medium-sized firms in developing new, innovative products and processes.[6] But there is no national technology delivery system that effectively bridges the public-private sector interface. Attempts have been made in the past to establish a national network along the lines of the agricultural extension service, but with little success. Evidently, it is not sufficient to stress technology availability or to build interface networks without also creating the appropriate conditions to stimulate technology demand in the private sector. Government initiatives for this purpose may take a wide variety of forms. New relationships between government and industry may have to be forged to strengthen competitive demand for technological advantage. These relationships will in all likelihood differ for different industrial sectors as the prevailing market factors and competitive conditions are generally sector specific.[5] It may be necessary to reexamine and clarify anti-trust legislation in favor of strengthening the competitive posture of American industry in international markets. Questions of monopoly and oligopoly should perhaps be resolved using flexible administrative procedures rather than adversarial confrontations based on rigid, costly legal actions. The disincentive to technological innovation caused by regulatory standards could be tempered by greater government assistance and counseling efforts to reduce the costs of compliance. Moreover, a whole range of fiscal and regulatory incentives could be instituted that would brighten the attraction for private enterprises to seek and utilize government technologies.

The system of tax and regulatory incentives available to American industry is a mere shadow of corresponding incentives available to German and Japanese industry. Considerably stronger U.S. incentives are needed to encourage industrial activity at all stages of the innovation process from inventive activity through R&D to marketing and commercialization. The promise of significant tax benefits and other incentives for undertaking high-risk, innovative projects is especially important to small, technology-oriented companies, which traditionally experience great difficulty in raising the risk-capital required. The ability of these firms to gain access to capital markets determines to a large degree the level of American innovativeness and competitiveness. Capital formation therefore is a key consideration in the innovation equation deserving of the closest attention by government legislators and policymakers. Greater incentives are needed to make risk capital more easily accessible to small- and medium-sized companies. A system of government credits and guarantees could be introduced to reduce financial risks for investors and venture-capital firms in backing risky, technologically oriented projects. Additionally, adequate attention should be devoted to promoting the creation of capital resources. In particular, the accumulation of personal savings must be made more attractive to the individual worker by perhaps raising the maximum interest rate obtainable from personal savings accounts. Alternatively or additionally, tax benefits could be introduced that would raise the effective rate of return possible from savings accounts. More disciplined savings accumulation would have a powerful, double-barreled effect on the economy in that it would increase the available supply of investment capital and at the same time help control inflation by curtailing consumer demand.

Another major factor concerns the American dedication to the work ethic. The tremendous growth of social welfare programs in the United States is mirrored by the fact that HEW, one of the newer cabinet-level departments on the Washington scene, also commands the largest annual budget of any federal agency. The danger in the proliferation of welfare programs is that it tends to erode the individual's desire to resume a productive work life. It contributes to the unhealthy attitude that society is responsible for the individual's daily sustenance. The need for social welfare is not being questioned as there are many legitimate recipients genuinely in need of government assistance. What does require attention is whether existing welfare programs are designed to assist recipients eventually to get off welfare rolls and to resume a productive work life. The main thrust of social welfare should be directed towards reeducation and retraining of the displaced and unemployed as a step towards reemployment instead of simply guaranteeing minimum incomes and distributing foodstamps. Retraining and reconstitution of the labor force is especially important at a time when restructuring of American industry is increasingly necessary to meet the foreign competitive challenge. A vigorous, industrious work force is an absolute necessity to undergird the revitalization of the nation's economy.

But the future of America rests on its youth. For this reason the nation's educational institutions will continue to play an important role in molding and preparing America's future leaders. Major tasks are to impart to students a basic understanding of the pervasive importance of technology in their lives and to instill in them a readiness to help direct technological change towards greater social and economic progress. In this respect educational institutions must work towards changing not-invented-here attitudes in individuals that lead one to resist the application of technology developed by others. A more balanced professional value system is needed that will assign as much weight to innovations based on acquired technology as is given to original development and scientific achievement. More interdisciplinary curricula are desired to help broaden individual perpectives and innovative capacities. Curricula designed to stimulate the innovational and entrepreneurial spirit of the individual, such as those tested by the NSF at the several university innovation centers, could be more widely adopted by educational institutions across the land.[7] In time such a reorientation by the nation's educational system could well prove the keystone by which substantive improvement in the U.S. technological innovation rate might eventually be achieved.

The foregoing discussion points to a number of important areas fundamental to America's future prosperity and growth. Drawn from seasoned observations of the American economic, social, and political scene and judiciously leavened with lessons from Japan and West Germany, they represent fundamental areas of institutional change affecting all sectors of American society. As such they provide the basis for prescribing improvements in the economic and social fabric of America. Recognition of the major issues and areas of change is absolutely essential to better position the nation for a future world that promises to be even more competitive and dynamic.

References

Chapter 1

1. "A National Policy for Productivity Improvement," National Commission on Productivity and Work Quality, Oct. 1975.
2. E. Ginzberg, *Technology and Social Change*, Columbia University Press, New York, 1964.
3. "Productivity I—Defining the Games and Players," *IEEE Spectrum*, **15**(10) (Oct. 1978): 34–52.
4. D. F. Koeppe, *Technology Transfer—Theory and Applications*, Twenty-First Century Corporation, Texas, 1977.
5. "Making U.S. Technology More Competitive," *Business Week* (Jan. 15, 1972): 46.
6. Edward B. Roberts and Alan L. Frohman, "Strategies for Improving Research Utilization," *Technology Review*, **80**(March/April 1978): 33–39.
7. Ernst R. Berndt and Dale W. Jorgenson, "How Energy, and Its Cost, Enter the 'Productivity Equation,' " *IEEE Spectrum*, **15**(10) (Oct. 1978): 50–52.
8. Sherman Gee, "The Role of Technology Transfer in Innovation," *Research Management*, **17**(6) (Nov. 1974): 31–36.
9. Sumner Myers and Eldon E. Sweezy, "Why Innovations Fail," *Technology Review* (March/April 1978): 41–46.
10. J.M. Utterback, "Innovation in Industry and the Diffusion of Technology," *Science*, **183**(Feb. 15, 1974): 620–626.
11. J. Brooks, *Telephone: The First Hundred Years*, Harper & Row, New York, 1975.
12. S.F. McKay, "The High-Technology Trap: Product Preoccupation—A Case in Point: the Laser Industry," *IEEE Transactions on Engineering Management*, **EM-19** (Feb. 1972): 31–32.
13. E.A. Haeffner, "The Innovation Process," *Technology Review*, **58**(March/April 1973): 18–25.
14. C.W. Sherwin and R.S. Isenson, "Project Hindsight," *Science*, **156**(June 23, 1967): 1571–1577.

213

15. R. Johnston and M. Gibbons, "Characteristics of Information Usage in Technological Innovation," *IEEE Transactions on Engineering Management,* **EM-22**(Feb. 1975): 27–34.

16. P. Jervis, "Innovation and Technology Transfer—The Roles and Characteristics of Individuals," *IEEE Transactions on Engineering Management,* **EM-22**(Feb. 1975): 19–27.

17. P.E. Connor, "Scientific Research Competence as a Function of Creative Ability," *IEEE Transactions on Engineering Management,* **EM-21** (Feb. 1974): 2–8.

18. Roy Rothwell, "Successful and Unsuccessful Innovators," *Planned Innovation* (Apr. 1979): 126–128.

19. "Barriers to Innovation in Industry," National Science Foundation report prepared by Arthur D. Little Inc. and Industrial Research Institute, Sep. 1973.

20. "Decade-Long Regulation vs Inflation Fight Continues," *Washington Post,* (Jan. 2, 1979): 10.

21. Frank Press, "Towards New National Policies to Increase Industrial Innovation," *Research Management,* **21** (Jul. 1978): 10–13.

22. "Polls are Consistent in Backing Environmental Cleanup Plans," *Washington Post* (Jan. 2, 1979): D11.

23. Kenneth E. Knight, George Kozmetsky, and Helen R. Baca, "Industry Views of the Role of the Federal Government in Industrial Innovation," unpublished report, The University of Texas at Austin, Jan. 1976.

24. S. Feinman and W. Fuentevilla, "Indicators of International Trends in Technological Innovation," Final Report, Gellman Research Association, Inc., Jenkintown, Pa., 1976.

25. Sherman Gee, "Factors Affecting the Innovation Time-Period," *Research Management,* **21**(1) (Jan. 1978): 37–42.

26. Edward L. Ginzton, "The $100 Idea," *IEEE Spectrum,* **12** (Feb. 1975): 30–39.

27. Christopher S. Derganc, "Thomas Edison and His Electric Lighting System," *IEEE Spectrum,* **16**(2) (Feb. 1979): 50–59.

Chapter 2

1. Keith Pavitt, "International Technology and the U.S. Economy: Is There A Problem?," in Rolf R. Piekarz, Ed., *The Effects of International Technology Transfers on the U.S. Economy,* National Science Foundation, Washington, D.C., 1973.

2. "Public Policy and Technology Transfer: Viewpoints of U.S. Business, Volume 1—Overview and Policy Considerations." Report prepared for the Department of State, Washington, D.C., March 1978.

3. "Science Indicators 1976." Report of the National Science Board, Washington, D.C., 1977.

4. "Science Indicators 1978." Report of the National Science Board, Washington, D.C., 1979.

5. "Technology Transfer from Foreign Direct Investment in the United States," The National Research Council, Washington, D.C., 1976.

6. W. Halder Fisher, "Technology Transfer as a Motivation for United States Direct Investment by European Firms." Report prepared for the National Science Foundation, Washington, D.C., Dec. 31, 1977

7. Regina K. Kelly, "The Impact of Technological Innovation on International Trade Patterns." Staff Economic Report ER-24, Bureau of International Economic Policy and Research, Department of Commerce, Washington D.C., Dec. 1977.

8. Regina K. Kelly, "Recent Trends in Technology Intensive Trade," private communication.

9. "The Multinational Goes International," *Technology Review* **81**(Dec. 78/Jan. 79):28.

10. M. Boretsky, "Export of U.S. Technology," *Management of Science and Technology,* Industrial College of the Armed Forces, Washington, D.C., 1976, pp. 63–67.

11. "Selected Trade and Economics Data of the Centrally Planned Economies," Bureau of East-West Trade, Department of Commerce, Washington, D.C., May 1975.

12. "The National Paper of the United States." Report prepared for the 1979 United Nations Conference on Science and Technology for Development, Vienna, Austria, Dec. 1978.

13. *The Technology Transfer Times,* **2**(12) (Dec. 1978).

14. J.B. Bingham and V.C. Johnson, "A Rational Approach to Export Controls," *Foreign Affairs,* **57**(Spring 1979) 894–920.

15. William A. Root, "Controls on West-to-East Technology Transfer," *Defense Systems Management Review,* **2**(1) (Winter 1979): 44–52.

16. "NATO Seeking Cohesive Armaments Policy," *Defense Electronics,* **11**(5) (May 1979): 58–77.

17. "PRC Arms," *Electronic Warfare/Defense Electronics,* **11**(1) (Jan. 1979): 52.

18. "An Analysis of Export Control of U.S. Technology—A DOD Perspective." Report of the Defense Science Board Task Force on Export of U.S. Technology, Office of the Director of Defense Research and Engineering, Washington, D.C., Feb. 4, 1976.

Chapter 3

1. "FY 1980 DOD Program for Research, Development and Acquisition," Statement by the Under Secretary of Defense for Research and Engineering to the 1st Session, 96th Congress, Feb. 1979.

2. "Technology Assessment and Forecast." Eighth Report, U.S. Patent and Trademark Office, Department of Commerce, Washington, D.C., Dec. 1977.

3. "Science Indicators 1978." Report of the National Science Board, Washington, D.C., 1979.

4. Benson Soffer, "Patent Activity and International Competitiveness," *Research Management,* **21**(Nov. 1978): 34–37.

5. Sherman Gee, "Factors Affecting the Innovation Time-Period," *Research Management,* **20**(1) (Jan. 1978) 37–42.

6. Jorge Miedzinski, "Technology Transfer in Canada," in Sherman Gee, Ed., *Technology Transfer in Industrialized Countries,* Sijthoff & Noordhoff International Publishers, Alphen aan den Rijn, The Netherlands, 1979, pp. 37–50.

7. Sherman Gee, "Questionnaire Results: A Comparative Assessment," in S. Gee, Ed., *Technology Transfer in Industrialized Countries,* Sijthoff & Noordhoff International Publishers, Alphen aan den Rijn, The Netherlands, 1979, pp. 433–437.

8. "Science Indicators 1976." Report of the National Science Board, Washington, D.C., 1977.

9. Richard S. Morse, "The Role of New Technical Enterprises in the U.S. Economy," Commerce Technical Advisory Board, Department of Commerce, Washington, D.C., 1975.

10. "Innovation—Has America Lost Its Edge," *Newsweek* (June 4, 1979): 67.

11. Harvey Brooks, "What's Happening to the U.S. Lead in Technology?," *Harvard Business Review,* **50** (May/June 1972): 110–118.

12. Edgar Weinberg, "A Call for Focusing on Productivity," *IEEE Spectrum,* **15**(10) (Oct. 1978): 34–39.

13. William Bowen, "Better Prospects for Our Ailing Productivity," *Fortune,* **100**(11) (Dec. 3, 1979): 68–86.

14. "Investment, Productivity, and Growth in Major Industrialized Countries," *Review of Economic and Financial Developments,* Department of the Treasury, Washington, D.C., Mar. 21, 1975.

15. "International Economic Report of the President," Council on International Economic Policy, Washington, D.C., Mar. 1976, pp. 59–63.

16. United Nations International Statistics Yearbook, 1966–1977.

17. *Commerce America* (June 19, 1978): 9.

18. Herbert E. Meyer, "Those Worrisome Technology Exports," *Fortune,* **97**(10) (May 22, 1978): 106–109.

19. "Siemens Starts Second But Finishes First," *Fortune,* **97**(7) (Apr. 10, 1978): 59–69.

20. Gene Gregory, "Japan Turns Its Ingenuity to the World Computer Market," *IEEE Spectrum* (Apr. 1979): 69–71.

21. Ted Szulc, "Trade War: The First Skirmishes," *Forbes* (Oct. 15, 1977): 29–30.

22. "Looming Threat from the ADCs," *Dun's Review* (Mar. 1979): 88–96.

23. Ezra Vogel, "The Miracle of Japan—How the Post-War Was Won," *Saturday Review* (May 26, 1979): 18–23.

24. "U.S.-Japanese Trade: The Awful Truth," *Forbes* (Mar. 19, 1979): 33–34.

25. "Why the Tokyo Round Was a U.S. Victory," *Fortune,* **99**(10) (May 21, 1979): 130–135.

26. "GATT Pact: Boon for Business," *Dun's Review,* (June 1979): 58–64.

27. "U.S. Companies in Unequal Combat," *Fortune,* **99**(7) (Apr. 9, 1979): 102–110.

28. "The Department of Defense Written Statement on NATO-Improved Armaments Cooperation," statement by the Under Secretary of Defense for Research and Engineering to the First Session, Ninety-sixth Congress, April 4, 1979.

29. Thomas A. Campobasso, "Market Imperatives of NATO Cooperation," *Signal* (Apr. 1979): 37–39.

30. Thomas A. Callaghan, Jr., "U.S./European Cooperation in Military and Civil Technology," Georgetown University, Washington, D.C., Sep. 1975.

31. Statement by General Andrew J. Goodpaster, Supreme Allied Commander, Europe (1969–1974).

32. Paul H. Backus, "NATO, EW, and Munitions Control," *Journal of Electronic Defense,* **2**(3) (May/June 1979). 21–63.

33. Daniel S. Greenberg, "Europe: Closing the Technology Gap," *Washington Post* (June 19, 1979).

34. "NATO Standardization, Interoperability and Readiness." Special Subcommittee Report of the House Armed Services Committee, Second Session, Ninety-fifth Congress, Washington, D.C., 1978.

35. Bernard Udis, "Technology Transfer in the Case of the F-16 Military Aircraft: A Preliminary Evaluation," in Sherman Gee, Ed., *Technology Transfer in Industrialized Countries,* Sijthoff and Noordhoff International Publishers, Alphen aan den Rijn, The Netherlands, 1979, pp.245–258.

36. "Allied Partnership in Armaments." Seminar proceedings, Georgetown University, Washington, D.C., Mar, 1977, pp. 28–31.

37. Robert N. Parker, "Technology Exchange with the USSR: National Security Issues," *Research Management,* (July 1974): 12–13.

Chapter 4

1. Daniel S. Greenberg, "Our Indolent Pursuit of Foreign Technology," *Washington Post* (April 11, 1978).

2. John W. Kiser III, "Report on the Potential for Technology Transfer from the Soviet Union to the United States." Report prepared for the Department of State and the National Science Foundation, Washington, D.C., Oct. 1977.

3. "Making U.S. Technology More Competitive," *Business Week* (Jan. 15, 1972): 44–49.

4. Betsy Ancker-Johnson and David B. Chang, "U.S. Technology Policy: A Draft Study," Department of Commerce, Washington, D.C., Mar. 1977.

5. Paul H. Backus, "NATO, EW, and Munitions Control," *Journal of Electronic Defense,* **2**(3) (May/June 1979): 21–63.

6. Dean Smith, National Technical Information Service, Springfield, Va., private communication, April 27, 1979.

7. George Krambles, "Public Transportation in Japan: Contrasts and Conclusions," *Transit Journal* (Aug. 1975): 29–38.

8. Koji Kobayashi, "Technology Transfer in Japan: An Introduction into the Current State of Japanese Industry," in Sherman Gee, Ed., *Technology Transfer in Industrialized Countries,* Sijthoff and Noordhoff International Publishers, Alphen aan den Rijn, The Netherlands, 1979, pp. 27–35.

9. K. Hirona, "Japan—Technology and R&D," *Unit 3,* Dvorkovitz & Associates, Ormond Beach, Fla., Apr. 1976, p. 13.

10. "Policies for the Stimulation of Industrial Innovation." Country Reports, Vol. 2-1, Organization for Economic Cooperation and Development, Paris, France, 1978, p. 341.

11. Gene Gregory, "Why Japan Succeeds," *IEEE Spectrum* (Mar. 1974): 65–72.

12. M. Uenohara in Sherman Gee, Ed., *Technology Transfer in Industrialized Countries,* Sijthoff and Noordhoff International Publishers, Alphen aan den Rijn, The Netherlands, 1979, p. 36.

13. Gene Bylinsky, "The Japanese Spies in Silicon Valley," *Fortune,* **97**(4) (Feb. 27, 1978): 74–79.

14. Gene Gregory, "Japan Turns Its Ingenuity to the World Computer Market," *IEEE Spectrum* (Apr. 1979): 69–71.

15. *Business Week* (Mar. 24, 1973): 57.

16. Ezra Vogel, "The Miracle of Japan—How the Post-War Was Won," *Saturday Review* (May 26, 1979): 18–23.

17. "Innovation—Has America Lost Its Edge?," *Newsweek* (June 4, 1979): 58–78.

18. "Science Indicators 1976." Report of the National Science Board, Washington, D.C., 1977.

19. Sherman Gee, "Factors Affecting the Innovation Time-Period," *Research Management,* **21**(1) (Jan. 1978): 37–42.

20. D.M. Collier, "Research-Based Venture Companies: The Link Between Market and Technology," *Research Management,* **17** (May 1974): 16–20.

21. Jack W. Pearson, "New Ways to Bring Technology to the Marketplace," *Technology Review,* **79**(5) (Mar./Apr. 1977): 26–35.

22. "Draft Report on Information Policy." Draft Report of the Subcommittee on Patent and Information Policy of the Advisory Committee on Industrial Innovation, Department of Commerce, Washington, D.C., Dec. 20, 1978.

23. "Foreign Investors in the U.S.—The Pace Quickens," *Forbes* (Apr. 2, 1979): 73.

24. "It Pays to Brave the New World," *Fortune,* **100**(2) (July 30, 1979): 86–91.

25. Lawrence Maloney, "Why Foreign Investors Put Their Money on U.S.," *US News & World Report* (July 9, 1979): 29–32.

Chapter 5

1. "Alternative Energy Sources," *Barrons,* (Sep. 17, 1979): 19.
2. "Energy Efficiency, European Style," *Technology Review* (Oct./Nov. 1976): 24.
3. "But Why Are German Businessmen Nervous," *Forbes,* **116**(2) (July 15, 1975): 18–25.
4. "Technology Enhancement Programs in Five Foreign Countries." Report 72-11412, U.S. Department of Commerce, Washington, D.C., Dec. 1972.
5. "Policies for the Stimulation of Industrial Innovation." Country Reports, Vol. 2-1, Organization for Economic Cooperation and Development, Paris, France, 1978.
6. Willi R. Steckelberg, "Compensating Employed Inventors in Europe," *Research Management,* **22**(4) (July 1979): 28–31.
7. Ezra Vogel, "The Miracle of Japan—How the Post-War Was Won," *Saturday Review* (May 26, 1979): 18–23.
8. "Japan," *Pan Am Clipper* (July 1975). 24-25.
9. Gene Gregory, "Why Japan Succeeds," *IEEE Spectrum,* **11** (March 1974): 65–72.
10. J. Herbert Hollomon, "Government and the Innovation Process," *Technology Review,* **81**(6) (May 1979): 30–41.
11. Jorge Miedzinski, "Technology Transfer in Canada," in Sherman Gee, Ed., *Technology Transfer in Industrialized Countries,* Sijthoff & Noordhoff International Publishers, Alphen aan den Rijn, The Netherlands, 1979, p. 39.
12. Bernard Udis, "Adjustment of High Technology Organizations to Reduced Military Spending: The Western European Experience." Report NSF-RA-X-75-017, National Science Foundation, Washington, D.C., Oct. 1974.
13. "Science Indicators 1978." Report of the National Science Board, National Science Foundation, Washington, D.C., 1979, p. 7.

Chapter 6

1. "U.S. Innovation: It's Better Than You Think," *Dun's Review* (Mar. 1979): 55–58.
2. Elmer B. Staats, "Technology Innovation: Improving the Climate for Government-Industry Cooperation." Address to the Industrial Research Institute, Boca Raton, Fla., May 11, 1976.
3. "How the Feds Will Rescue Innovation," *Technology Review* (Oct. 1979): 74–75.
4. Frank O. Ohlson, "Trends in Federal Patent Policy," *Defense Systems Management Review,* **2**(1) (Winter 1979): 10–15.

5. Betsy Ancker-Johnson and David B. Chang, "U.S. Technology Policy: A Draft Study," Department of Commerce, Washington, D.C., Mar. 1977.

6. Sherman Gee, Ed., *Technology Transfer in Industrialized Countries,* Sijthoff and Noordhoff International Publishers, Alphen aan den Rijn, The Netherlands, 1979, pp. 364–369.

7. Sherman Gee, "Military-Civilian Technology Transfer: Progress and Prospects," *Defense Management Journal,* 11(2) (Apr. 1975): 46–51.

8. "Technology Transfer and Utilization." Report of the National Academy of Engineering, Washington, D.C., 1974, p. i.

9. "Means for Increasing the Use of Defense Technology for Urgent Public Problems." General Accounting Office report to Congress, Washington, D.C., Dec. 1972.

10. "Intergovernmental Use of Federal R&D Laboratories," Federal Council for Science and Technology, Washington, D.C., Mar. 1974.

11. Frederick L. Haynes, "Stimulating Innovation Through U.S. Policy: A Cooperative Technology View," in Conference Proceedings, 1979 IEEE USAB Conference on U.S. Technology Policy, The Institute of Electrical and Electronics Engineers, Inc., New York, May 1–3, 1979, pp. 40–44.

12. Fred A. Tarpley, "Domestic Policy Review of Industrial Innovations and Recommendations for Creating Jobs Through the Success of Small Innovative Business," in Conference Proceedings, 1979 IEEE USAB Conference on U.S. Technology Policy, May 1–3, 1979, pp. 47–50.

13. Sherman Abramson, "Recommendations for Creating Jobs Through the Success of Small Innovative Businesses," in Conference Proceedings, 1979 IEEE USAB Conference on U.S. Technology Policy, May 1–3, 1979, pp. 39–42.

14. "Small Business and Innovation," Office of Chief Counsel for Advocacy, Small Business Administration, Washington, D.C., May 1979.

15. "Science Indicators 1976," Report of the National Science Board, Washington, D.C., 1977

16. "Policies for the Stimulation of Industrial Innovation." Country Reports, Vol. 2-1, Organization for Economic Cooperation and Development, Paris, France, 1978.

17. "Technological Innovation and Federal Government Policy-Research and Analysis." Report NSF 76-9, Office of National R&D Assessment, National Science Foundation, Washington, D.C., Jan. 1976.

18. "Program Announcement for Extramural Research," Division of Policy Research and Analysis, National Science Foundation, Washington, D.C., Dec. 1978.

19. "Federal Technology Transfer," Federal Coordinating Council for Science, Engineering, and Technology, June 1977, pp. 1–8.

20. James A. Higgins, "Technology Transfer: A Key to Productivity," *Defense Systems Management Review,* 2(1) (Winter 1979): 7–9.

21. Robert C. Crawford, "The Federal Laboratory Consortium for Technology Transfer," *Defense Systems Management Review,* **2**(1) (Winter 1979): 53–64.

22. *Machining Briefs* (Mar./Apr. 1974).

23. Roland T. Tibbetts, "NSFs Three Phase Program Helps the Small-Business Innovator Bootstrap An Idea to Commercial Success," *IEEE Spectrum,* **15**(10) 86(Oct. 1978).

24. Robert M. Colton, "Technological Innovation Through Entrepreneurship," in Sherman Gee, Ed., *Technology Transfer in Industrialized Countries,* Sijthoff and Noordhoff International Publishers, Alphen aan den Rijn, The Netherlands, 1979, pp. 173–193.

25. Albert H. Teich and Melvin M. Whartnaby, "Technology Transfer and Military Research: The U.S. Naval Weapons Center and the Federal Laboratory Consortium." Report prepared for the National Science Foundation, Washington, D.C., undated.

26. James E. Jarrett and Benjamin J. Jones, "Creating Jobs: Connecticut's Product Development Corp.," Council of State Governments, Lexington, Kentucky, Dec. 1978.

27. John N. Phillips, Connecticut Product Development Corp., Hartford, Connecticut, private communication, Aug. 1979.

28. "Innovation—Has America Lost Its Edge," *Newsweek* (June 4, 1979): 63–68.

29. "Report on Government Patent Policy," Federal Council for Science and Technology, Washington, D.C., Sep. 1976, pp. 401–405.

30. David T. Mowry, "Patent Management at NTIS," *Defense Systems Management Review,* **2**(1) (Winter 1979): 22–26.

31. "The Patent System Needs Overhauling," *Business Week* (Dec. 4, 1971): 65–68.

32. Roger M. Milgrim, "Get the Most Out of Your Trade Secrets," *Harvard Business Review,* **52**(Nov./Dec. 1974): 105 112.

Chapter 7

1. "Technology Enhancement Programs in Five Foreign Countries." Report COM-72-11412, Department of Commerce, Washington, D.C., 1972, p. 240.

2. "Policies for the Stimulation of Industrial Innovation." Country Reports, Vol. 2-1, Organization for Economic Cooperation and Development, Paris, France, 1978, p. 311.

3. Sherman Gee, Ed., *Technology Transfer in Industrialized Countries,* Sijthoff and Noordhoff International Publishers, Alphen aan den Rijn, The Netherlands, 1979, p. 373.

4. Ezra Vogel, "The Miracle of Japan—How the Post-War Was Won," *Saturday Review* (May 26, 1979): 23.

5. J. Herbert Hollomon, "Government and the Innovation Process," *Technology Review,* **81**(6) (May 1979): 30–41.

6. Glenn B. Fatzinger and Sherman Gee, "A Technology Network for Economic Development," *Journal of Technology Transfer,* **4**(2) (Spring 1980).

7. Robert M. Colton, "Technological Innovation Through Entrepreneurship," in Sherman Gee, Ed., *Technology Transfer in Industrialized Countries,* Sijthoff and Noordhoff International Publishers, Alphen aan den Rijn, The Netherlands, 1979, pp. 173–190.

Index